BASIC PROCESS MEASUREMENTS

BASIC PROCESS MEASUREMENTS

Cecil L. Smith

WILEY

A JOHN WILEY & SONS, INC., PUBLICATION

Published by John Wiley & Sons, Inc., Hoboken, New Jersey
Published simultaneously in Canada

For general information on our other products and services or for technical support, please contact our Customer Care Department within the United States at (800) 762-2974, outside the United States at (317) 572-3993 or fax (317) 572-4002.

Wiley also publishes its books in a variety of electronic formats. Some content that appears in print may not be available in electronic formats. For more information about Wiley products, visit our web site at www.wiley.com

ISBN: 978-0-470-38024-6

Library of Congress Cataloging-in-Publication Data is available.

Printed in the United States of America

10 9 8 7 6 5 4 3 2 1

■ CONTENTS

My business is exclusively process control. And like many other activities, process control relies on measurement devices—every control loop contains at least one measurement device that provides a very critical function. Most activities, and especially process control, are subject to the garbage-in, garbage-out characterization. The reason is simple—decisions based on bad data are likely to be bad decisions. The Iraq War is a prime example.

This book is intended for anyone involved in the application of measurement devices in an industrial environment. The target audience includes chemical engineers, mechanical engineers, electrical engineers, and industrial chemists, but anyone with a technical background will find this book helpful for identifying appropriate measurement devices for an application. For the benefit of those with a limited process background, the relevant basic concepts are explained at the beginning of each chapter. For example, the Reynolds number is explained at the beginning of Chapter 5, which covers flow measurement. Chemical engineers and mechanical engineers are certainly familiar with the Reynolds number, but most electrical engineers and industrial chemists are far less familiar, if at all.

This book concentrates on measurement devices for the basic variables: temperature, pressure, level, density, and flow. This book does not attempt to cover every possible approach to measuring these variables but instead focuses on technologies that are most commonly installed in industrial facilities. Equal emphasis is given to the attributes that make each attractive and to the factors that lead to problems. One must understand both the process and the principles on which the measurement device relies. Remember, those who know what they are doing get what they pay for; those that do not get what they deserve!

When it comes to process operations, measurement devices have become our eyes. There was a time when process operators could rely on their own senses to make decisions. In the paper industry, they could reach into a stock tank, grab a handful of the fiber, squeeze out the water, use the sole of their shoe to form a crude mat or sheet, and then tell you what kind of paper it would make. Perhaps the most impressive was a person one who could chew a polymer sample for a few minutes and tell you more than the QC lab could tell you hours later! Did his manager know he did this? Sometimes he did, and sometimes he did not.

But the way to reduce emissions is to replace open tanks with closed tanks and otherwise button up a process. We are correctly more sensitive to employee exposure to industrial chemicals, so things that were tolerated back in the 1960s when I got into this business are appropriately taboo today. I have also witnessed the adjustments, occasionally painful, that occurred when production operators were forced to rely increasingly more, and in some cases exclusively, on the information provided by the measurement devices.

For a commercial measurement device to be successful in process installations, two requirements must be satisfied:

- *The measurement device must rely on a sound basic principle.* In many cases, this basic principle establishes boundaries in which the measurement device can be used successfully. For example, the vortex shedding flow meter cannot be used to measure flows in the laminar region. The better one understands the principles behind a measurement device, the better one can recognize viable applications for the measurement device and avoid misapplications and the ensuing consequences, which occasionally extend beyond no return for the money and time invested.

- *The measurement device must be constructed in a manner consistent with the process conditions to which it will be exposed.* Designing a measurement device for industrial service is definitely a specialty. Unfortunately, some of the best in the business occasionally stumble. In the early efforts to use thin film technology in pressure measurement devices, the strain gauges were bonded to the pressure-sensitive diaphragm. In essence, *bonded* means "glued"; the use of this technology in industrial measurement devices has been largely unsuccessful.

In modern measurement devices, the basic principle usually involves the translation (by a sensor or transducer) of the process variable of interest (temperature, flow, etc.) to an electrical property (voltage, resistance, capacitance, etc.) that can be sensed. These are generally well understood, and most can be expressed mathematically and analyzed. But rarely is this the case for the manner in which the measurement device is constructed. One learns this from experience—that is, install a few of a given model and see how well they perform. Maintenance issues usually take the front seat, and the potential problems may not surface for several years.

Some industries have special requirements, such as sanitary conditions or food approval. But in the end, the requirements of the various segments of the process industries are more alike than different. The manufacturers respond by offering their products in different models to suit a variety of special requirements. For example, the filling fluid in a capillary seal system can potentially leak into the process. When you install such devices in plants that operate 24/7 for many years, anything that *can* happen *will* happen. The consequences of the filling fluid leaking into the process must be tolerable.

As noted previously, I get into process measurement issues primarily through my process control activities. When troubleshooting process control problems, you always have to include the possibility that a measurement device is lying to you, perhaps only under certain situations. With the incorporation of microprocessor technology into the measurement devices, this is becoming less frequent, but has not and probably never will entirely disappear. Incorporating the microprocessor within the measurement device improves the signal processing but also enhances the capability to detect when something has gone awry. As we replace current loop interfaces with digital communications, such situations can be more effectively reported.

In process control endeavors (and in others as well), it is imperative that we take advantage of new technologies and earnestly pursue continuing process improvement by:

- Resolving problems with currently installed measurement devices.
- Installing measurement devices whose performance is superior to those currently installed.
- Installing measurement devices for process variables that were not previously measurable.

Both of these have demonstrated the capability to produce recognizable improvements in process operations (improved product quality, better economic returns, etc.).

In process installations, we usually get it right for the normal process operating conditions. But every process inevitably operates, at least for short periods, under conditions very different from the normal process operating conditions. This is appropriately a consideration during the hazards analysis—that is, what will these operating conditions be? and Will all measurement devices perform properly under these conditions?

A major concern is a multiple failure accident, with one of the failures being that a measurement device that is lying to us. Unfortunately, presenting a bad piece of information during an abnormal event will likely compound the consequences. When you have only one problem, you usually recognize it quickly. But when you have two or more at the same time, it takes a little longer.

I am alarmed by the extent to which new developments are now coming from outside the United States. The DIN standard for the 100-Ω RTD came from Germany. Rosemount pioneered the capacitance cell for pressure transmitters, but Yamatake commercialized the piezoelectric technology and Yokogawa, the resonant frequency technology. Micro Motion pioneered the coriolis flow meter, but Khrone was first with a straight tube version. The Europeans clearly led the transition to intrinsically safe installations, using barriers from MTL (England) and R. Stahl (Germany). I am alarmed, but not surprised. With the emphasis on the financial sector, investing in technology and manufacturing has not been in vogue in the United States for some time.

In the latter stages of preparing this book, I chose to remove all reproductions of commercial products. These quickly become out of date (a couple of them were already out of date before I removed them). Today, there is a very easy way to obtain information on the latest models of commercial products for measurement devices: the Internet and a search engine! For example, if you want the latest on magnetic flow meters, just do a search. For the same reason, I have included the temperature–voltage relationship only for the type J thermocouple, and not even all of the available data. You can easily download the complete table for any type of thermocouple from the National Institutes of Science and Technology (NIST) website. Including such tables in a book makes no sense.

The process control business has been very good to me. As a consultant, I get to work on both diverse and interesting problems. However, I thoroughly enjoy teaching professional development courses, and I spend about a third of my life doing that. All are in some way related to process control, but included in my offerings is a course on process measurements.

Finally, a special thanks to my wife, Charlotte. She endures my staring into a computer screen for hours at a time. But fortunately I can now do this in places like Taos and Key West.

CECIL L. SMITH
Baton Rouge, Louisiana
November 2, 2008

CHAPTER 1

Basic Concepts

This chapter is devoted to topics that are common to all measurement devices.

Measurement devices can be characterized in several different ways. In regard to the measured value, some are continuous and some are discrete. In regard to time, some are continuous and some are sampled. In regard to their relationship to the process, some are in-line, some are on-line, and some are off-line.

The steady-state characteristics of a measurement device often determine its suitability for a given purpose. This includes its measurement range, its accuracy, its repeatability, the resolution of the measured value, and its turn-down ratio. Measurement uncertainty is receiving increasingly more attention and will probably receive even more in the future.

Most measurement devices provide values for functions performed by other systems, including data acquisition and process control. The older interfaces consisted largely of current loops. Although most microprocessor-based transmitters also provide a current loop output, the trend is to use network communications with field devices, a technology generically referred to as *fieldbus*. However, serial communication has not entirely disappeared.

The sensor portion of a measurement device is generally exposed to process temperatures, process pressures, etc. Considerations such as ambient temperature and the hazardous area classification apply to the transmitter part of the measurement device. Proper enclosures are required for every measurement device.

The dynamic characteristics of a measurement device are especially important in applications such as process controls. Lags—first-order lags or transportation lags (dead times)—may be associated with the measurement device. Filtering and smoothing the measured value result in additional lags, so these technologies must be applied very carefully so as to not degrade the performance of the process controls.

1.1. CONTINUOUS VS. DISCRETE MEASUREMENTS

Continuous and *discrete* refer to the type of value that is produced by the measurement device.

Continuous Measurements

The output of a continuous measurement device (often called a *transmitter*) indicates the current value of the variable being measured. The element *LT* (level transmitter) in Figure 1.1 represents a continuous measurement of the level in the evaporator. Provided the level is within the range covered by the measurement device, the output from the level transmitter indicates the level within the evaporator.

All continuous measurement devices are constrained by their measurement range, which in turn may be constrained by the technology, by the design parameters of the measurement device, by how the measurement device is connected to the process, and so on. The following terms pertain to the measurement range:

Lower-range value. Lower limit of the measurement range.

Upper-range value. Upper limit of the measurement range.

Span. Difference between the upper-range value and the lower-range value.

The output of a continuous measurement device is generally referred to as the *measured value* or *measured variable*. In process applications, the output of a continuous measurement device is often called the *process variable*.

Other examples of variables for which continuous measurements are available are temperature, pressure, flow, density, and composition. Parameters such

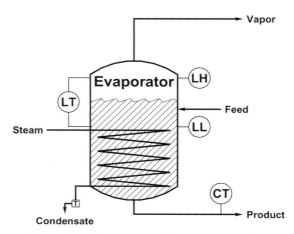

Figure 1.1. Measurements for an evaporator.

as accuracy, repeatability, turndown ratio, and resolution are associated with continuous measurements. These will be examined later in this chapter.

Discrete Measurements

The output of a discrete measurement device (often called a *switch*) is one of two states, depending on the value of the variable being measured. The elements *LH* (level high) and *LL* (level low) in Figure 1.1 represent discrete measurement devices in the form of level switches, one to detect that the liquid level is abnormally low and the other to detect that the liquid level is abnormally high.

The level switch basically indicates the presence or absence of liquid at a given point within the vessel, usually at the physical location of the switch. If the liquid level is above this location, the output of the level switch is one state. If the liquid level is below this location, the output of the level switch is the other state. In practice, there is always a small switching band (sometimes called the *deadband*) associated with a discrete measurement device.

The parameters associated with process switches are simpler than those associated with transmitters, and we will examine these parameters next.

Actuation and Reactuation   The state of the process switch changes when the appropriate conditions are present within the process. Consider the level switches within the evaporator in Figure 1.1. Each level switch changes state when the level within the evaporator attains the location of that level switch.

These level switches are said to actuate on rising level. Similarly, a pressure switch actuates on rising pressure. The behavior of a level switch is as follows (most other switches behave in a similar manner):

Actuation point. This is the vessel level at which the switch changes state on rising level. Sometimes the actuation point is referred to as the *set point.* The actuation point of most level switches cannot be adjusted; the actuation point is determined by the physical location of the level switch. However, many pressure switches provide an adjustable actuation point. When the actuation point of a pressure switch is adjustable, the working pressure is the range of pressures over which the actuation point can be specified.

Reactuation point. This is the vessel level at which the switch changes state on falling level. This occurs at a level below the actuation point. The *deadband* is the difference between the actuation point and the reactuation point. In most switches, the deadband is fixed, but it is occasionally adjustable. A few pressure switches provide separate adjustments for the actuation point and the reactuation point (or the actuation point and the deadband).

This behavior is often represented as a diagram (Fig. 1.2).

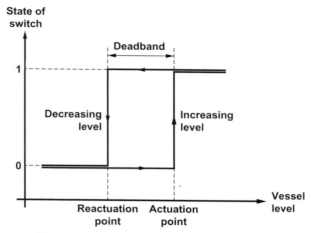

Figure 1.2. Switching logic for a level switch.

Normally Open and Normally Closed   The terms *normally open* and *normally closed* are basically equipment terms, not process terms. The normal state of a switch in no way implies that the corresponding process conditions are normal.

The normal state of a switch is its state at ambient conditions. Some authors refer to this as its *shelf state*, and this is a good way to think of the normal state of a switch. The normal state is the state of the switch when removed from the process and placed in the warehouse. For a level switch, the normal state would not indicate the presence of liquid.

Within each switch, there is a contact whose state can be sensed. Figure 1.3 shows several possible configurations:

- A single-pole, single-throw (SPST), normally open (NO) switch provides only one contact whose shelf state is open. A level switch of this type would be referred to as a *normally open level switch*. On actuation, this contact closes.

- An SPST normally closed (NC) switch provides only one contact whose shelf state is closed. A level switch of this type would be referred to as a *normally closed level switch*. On actuation, this contact opens.

- A single-pole, double-throw (SPDT), switch provides a contact for both states of the switch. As illustrated in Figure 1.3, there are three wiring connections:

    Common. This is the return or ground for the electrical circuit.

    NO. In the normal, or shelf, state for the switch, this contact is open (or no continuity between the NO terminal and the common terminal). On actuation, this switch closes.

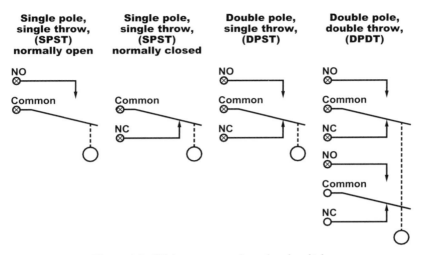

Figure 1.3. Wiring contacts for a level switch.

NC. In the normal, or shelf, state for the switch, this contact is closed (or continuity between the NC terminal and the common terminal). On actuation, this switch opens.

- A double pole, double throw (DPDT) switch is basically two double throw switches driven by the same mechanism. The switch provides six wiring connections.

The simple switches are single throw and must be ordered as either normally open or normally closed (the option is usually but not always available). Most switches designed for process applications are double pole and can be wired to either the normally open contact or the normally closed contact.

Wiring Diagram Symbols   The symbols used in the wiring diagrams reflect the type of switch (level, pressure, etc.) and specify which contact (NO or NC) is used. Figure 1.4 provides symbols as commonly used in wiring diagrams. The symbols are provided as follows:

- Level switch that actuates on rising level.
- Pressure switch that actuates on rising pressure.
- Temperature switch that actuates on rising temperature.
- Flow switch that actuates on increasing flow.
- Physical contact, often referred to as a limit switch, that is used on two-position valves, on doors, etc.

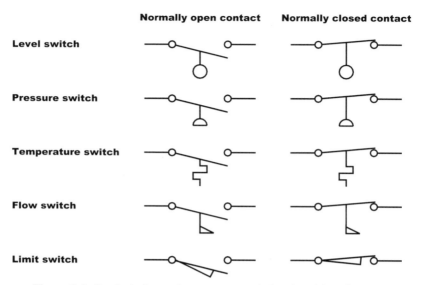

Figure 1.4. Symbols for various process switches in wiring diagrams.

Most companies have their preferred symbology for the various switches, but that given in Figure 1.4 is typical.

The required logic can be formulated using either the NC contact or the NO contact. The selection of which contact to use is determined by what conditions within the process are considered to be normal. The guiding principle is simple: When the process conditions are normal, all circuits that include a process switch must have continuity, which means that the wiring is to whichever contact (NO or NC) has continuity when the process conditions are normal. In this way, any failure in the circuit will indicate a problem. Otherwise, a defect in the circuit could easily go undetected and abnormal conditions would not be indicated.

Discrete Logic   Process switches often provide the inputs to safety and shutdown systems. For the evaporator in Figure 1.1, the steam is to be blocked in either of the following conditions:

> *Low level.* With an abnormally low level, the upper part of the tube bundle would not be submerged in liquid, which usually results in scaling or some other detrimental effect on the heat transfer surface.
> *High level.* With an abnormally high level, liquid would be entrained into the vapor stream exiting the evaporator (many evaporators have a mist extractor in the top that must not be partially submerged in the liquid).

Another way of viewing this is to state the conditions or permissives that must be true for the steam block valve to open:

Evaporator LL switch indicates presence of liquid. Under this condition, the LL switch is actuated. The NO contact would be closed, thus providing continuity in an electrical circuit.

Evaporator LH switch does not indicate presence of liquid. Under this condition, the LH switch is not actuated. The NC contact would be closed, thus providing continuity in an electrical circuit.

Traditionally, such logic was implemented in hard-wired electrical circuits. Figure 1.5 presents the wiring diagram for a circuit that determines if the current process conditions permit the steam-block valve to be open. The *circle* represents a coil that indicates that it is okay for the steam-block valve to be open (contacts on this coil would be used in other circuits that open the steam-block valve). The coil is energized if power flows from the power rail to ground, passing through the coil. Power will flow if the normally open contact on the level low switch is closed and the normally closed contact on the level high switch is closed. Continuity is required for the steam-block valve to be open.

Today, such logic is more likely to be implemented in programmable electronic systems such as programmable logic controllers (PLCs). At least in the United States, the representation of this logic is usually by relay ladder diagrams that are similar to the wiring diagrams used for hard-wired implementations. However, this is discrete logic and can be represented and implemented in a number of ways, including Boolean expressions and sequential function charts.

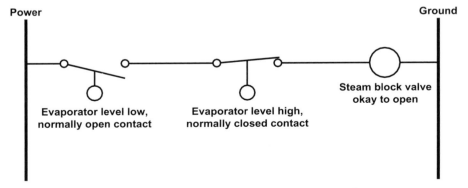

Figure 1.5. Permissive logic for a steam-block valve.

1.2. CONTINUOUS VS. SAMPLED MEASUREMENT

Continuous measurement and *sampled measurement* refer to the frequency with which the value from the measurement device is updated to reflect current conditions within the process.

Continuous Measurements

At every instant of time, a continuous measurement device provides a value that reflects the current value of the variable being measured. The level transmitter and both level switches in Figure 1.1 are continuous in this sense. Measurements for temperature, pressure, level, and flow are almost always continuous from the perspective of time.

When microprocessors are incorporated into the measurement device, the output is updated very frequently but technically not continuously. With update rates such as 10 times per second, the result is equivalent (from a process perspective) to a continuous measurement device, and such devices are normally included in the continuous category.

Sampled Measurements

The element CT (composition transmitter) on the product stream from the evaporator in Figure 1.1 is possibly a sampled measurement. A few composition analyzers are continuous, but many involve sampling. A sample is withdrawn from the process, and the analysis is performed on this sample. Sometimes a complete composition analysis (consisting of several values) is generated; sometimes the measurement is reduced to a single value, such as the ratio of two key components or the total amount of impurities.

The *sampling time* is the time between analyses. The value for the sampling time depends on the complexity of the analysis. It may be as short as a few seconds, or it could be several minutes. Sometimes analyzers are multiplexed between several process streams, which extends the sampling time even further for a given measured variable.

When a *sample-and-hold* capability is incorporated, a value for the output is available at every instant of time. However, the output reflects the results of the most recent analysis and will not change until a new analysis is performed. Consequently, such measured values are still considered to be sampled and not continuous.

1.3. IN-LINE, ON-LINE, AND OFF-LINE

In-line, on-line, and *off-line* pertain to the physical relationship between the measurement device and the process. This discussion will also refer to the two categories of properties:

Intensive. An intensive property does not depend on the amount of material present. Intensive properties include temperature, composition, and physical properties. The values for such properties can be obtained by withdrawing a sample of material from the process and then analyzing for the desired value. For example, one could withdraw a sample and then determine its temperature. Although rare for temperature measurement, composition measurements are routinely done in this manner.

Extensive. An extensive property depends on the amount of material present. Extensive properties include flow, weight, and level. Measurements of these cannot be performed on a sample. For example, it is not possible to determine the flow through a pipe by withdrawing a sample.

In-Line Measurements

An in-line measurement is connected in such a manner that the measurement device directly senses the conditions within the process. Most basic measurements (temperature, pressure, level, and flow) are in-line. However, very few composition measurements are in-line.

In-line is the preferred approach. Unfortunately, this option is not always available. Occasionally in-line measurements are available, but they have major concerns that must be addressed. For example, the composition of the product from a caustic evaporator can be inferred from the density. One approach to measuring density is to use a nuclear density gauge. Such gauges sense the density of the material flowing in the product pipe and are thus in-line measurements. But before such gauges are installed, the issues associated with having radioactive materials on-site must be addressed.

In-line measurements can be classified as *contact* and *noncontact*. Noncontact measurement devices perform their functions without any contact to the contents of the process. Examples of noncontact measurement devices include the following

- Radiation devices for measuring level or density.
- Pyrometers for measuring temperature.
- Clamp-on versions of ultrasonic flow meters (mounted externally to the pipe containing the fluid).
- Microwave radar level measurement, sometimes referred to as *noncontact radar*. Because the antenna is mounted above the surface to be detected, it is noncontact with respect to the liquid or solids; however, the antenna is exposed to the gases and vapors present above the surface, which over time can lead to deposits and buildups in some applications.

Noncontact does not always provide total immunity from process considerations. Consider the clamp-on ultrasonic flow meters. The transmitters/receivers are not exposed to the process fluids, but they must be bonded to the outside surface of the pipe. Therefore, the transmitters/receivers are exposed to temperatures only slightly different from that of the process fluid.

In-line measurements are also classified as *intrusive* or *nonintrusive*. A nonintrusive measurement device performs its functions without affecting the activities within the process in any way. Examples of nonintrusive measurement devices include the following:

- A magnetic flow meter must contact the fluid but does not provide any obstruction to fluid flow. The flow tube of such a meter is basically a straight length of pipe that would have the same effect on fluid flow as a spool section of the same length. There are no regions of low pressure that could lead to flashing or cavitation.
- Pressure transmitters require a process connection but one that usually does not intrude into the process. When flush connections are required to avoid dead spaces, capillary seal arrangements can be installed.
- The antenna of a microwave radar transmitter normally extends slightly into the process vessel. Technically, this is intrusive but only to a nominal degree. If necessary, a plastic seal can be installed to completely separate the antenna from the process. But whereas the exposed antenna can be inserted through a nozzle as small as 2 in., a 12 in. or larger opening is required for the seal arrangements.

On-Line Measurements

On-line measurements sense the conditions of materials that are withdrawn from the process. The most common example of this is associated with composition analyzers. Figure 1.6 shows an installation in which a sample is withdrawn from the process, transported to the physical location of the analyzer, and then conditioned or cleaned up in some manner to provide material for analysis. From the perspective of time, the analyzer may be continuous (such as infrared or ultraviolet) or may be sampling (chromatograph).

Known as a *sample system*, the equipment for withdrawing material from the process, transporting it to the analyzer location, and then removing unwanted constituents tends to be a source of problems. The design of such systems requires special skills in handling small streams in an industrial environment. Any lapses lead to maintenance problems. Unfortunately, for analyzers such as chromatographs, there seems to be no other alternative.

Because they are intended for sampling, on-line measurements can be used only for intensive properties.

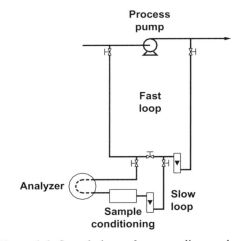

Figure 1.6. Sample loops for an on-line analyzer.

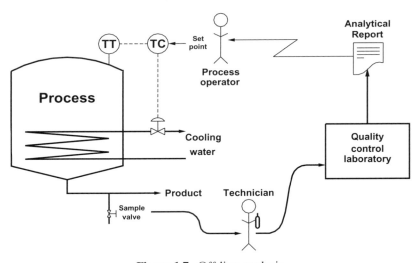

Figure 1.7. Off-line analysis.

Off-Line Measurements

Especially with the advent of robotics, almost any analysis could be implemented in an on-line fashion. But when the sampling interval is once every 4 or 8 hr, the cost justification becomes difficult. Furthermore, some very complex analyzers are much more suited to a laboratory environment (with lab technicians) than to a process environment (with process operators).

Figure 1.7 illustrates an off-line measurement. A sample is withdrawn from the process and transported to the laboratory, where the sample preparation is performed, the analysis is made, and the results are communicated back to the plant control room. From a control perspective, the major issue is how quickly the results are returned to the process operators. Technologies such as pneumatic conveying systems can speed the transport of the sample to the laboratory. Networking capabilities can make the results available in the control room as soon as they are obtained (and possibly verified) by the lab technicians.

Because they are intended for sampling, off-line measurements can be used only for intensive properties.

1.4. SIGNALS AND RESOLUTION

The output of a measurement device is generally referred to as a signal. In measurement applications, a signal is physical variable that in some way represents the process variable being measured. The signal is a mechanism for transfering information from one device (such as a measurement device) to another (such as a controller).

In process applications, the physical nature of the signal has evolved over the years and will probably continue to evolve. In the 1950s, most systems were pneumatic, using a 3- to 15-psi pneumatic signal. In the 1970s, electronic systems appeared, most using a 4- to 20-ma current loop to transmit information from one device to another. Both are analog transmission systems.

Eventually these will be replaced with digital communications using technology commonly referred to as fieldbus. These technologies permit several measured variables to be transmitted via a single physical conductor (a wire). The physical conductor can be eliminated by using technologies such as fiberoptics and wireless. However, all of these are digital communications systems.

We will later examine current loops and various forms of digital signal transmission systems, but not pneumatic systems.

Analog Values

In process applications, analog values are associated with signals such as the 3- to 15-psi pneumatic signal and the 4- to 20-ma current loop. Measurement devices such as a level transmitter can sense the level in the vessel, provided it is within the measurable range. The lower end of the measurable range corresponds to 3 psi or 4 ma; the upper end of the measurable range corresponds to 15 psi or 20 ma.

At least conceptually, an analog signal can represent any value of the vessel level, provided it is within the measurable range. In practice, there are limitations.

The resolution of the measurement device is the change in the process variable required to cause a change in the output signal. Basically, the resolution is the smallest change in the process variable that can be detected by the measurement device.

Digital Values

A digital value is an approximation to an analog value. The larger the number of digits in the digital value, the closer the approximation will be to the analog value. Computers represent data using the binary number system, and either a fixed-point or floating-point representation. But let's not get into the gory details of bits and bytes. We can illustrate our points using a fixed-point decimal representation.

Suppose the level in an evaporator can be measured over the range of 6 to 8 ft. Let's represent this as a digital value using a three-digit digital representation. The evaporator level is truly an analog value. Its conversion to the digital value is illustrated by the graph in Figure 1.8. The evaporator level is slowly increasing with a constant rate of change. However, the digital value increases in increments of 0.01 ft. Thus the minimum change that can be detected in this case is 0.01 ft. Consequently, the digital representation imposes a resolution of 0.01 ft.

In computer circles, the precision of a value is the number of digits following the decimal point. For example, the statement "Out.precision(2)" in C++ causes floating-point values written to stream Out to be formatted with two digits after the decimal point. When understood in this context, the precision of the digital value in Figure 1.8 would be 2. But within measurement circles, the term *precision* has been used (and misused) in so many ways that it is probably best to avoid it.

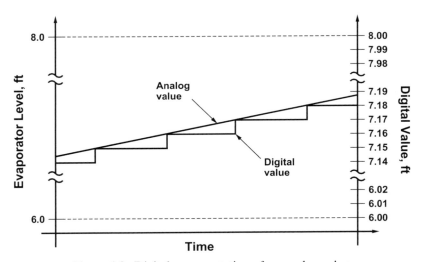

Figure 1.8. Digital representation of an analog value.

Resolution

When the measurement device provides a digital value, the best possible resolution is a change of 1 in the least significant digit of the digital value. However, it is not ensured that the measurement device can actually detect such a change. That is, the resolution can be no better than that imposed by the digital representation, but it can be worse.

The resolution can be expressed in engineering units (for example, 0.01 ft in Figure 1.8). Alternatively, it can be expressed relative to the measurable range (2.00 ft in Figure 1.8). On this basis, the resolution would be 0.01 ft in 2.00 ft, or 1 in 200 (usually read as *1 part in 200*). Most level measurement devices are capable of better resolution.

In determining the number of digits to be used in a digital value, the objective is for the resolution of the digital value to be sufficient to represent any meaningful change in the output of the measurement device. The specifications for most measurement devices state either the resolution or the repeatability (to be defined shortly). The resolution provided by the digital value should exceed both.

Voltage Inputs   Consider a measurement device that outputs a 0- to 5-V DC signal that represents its measured value. An analog to digital (A/D) converter within the data acquisition or control system converts this voltage to a digital value.

Suppose a pressure measurement device generates 0 V when the pressure is 0 mm Hg absolute and 5 V when the pressure is 800 mm Hg absolute. An A/D converter with a resolution of 1 part in 4000 would generate the following digital values:

Quantity	Lower Range Value	Upper Range Value
Absolute pressure (mm Hg)	0	800
Input (V DC)	0	5
Raw value (count)	0	4000

A change of 1 in the digital value is often referred to as a *count* and cannot have a fractional part. Therefore, the resolution of the A/D converter is 1 count, which corresponds to 5 V/4000 counts = 0.00125 V/count = 1.25 mV/count. In terms of the pressure, the resolution is 800 mm Hg/4000 counts = 0.2 mm Hg/count, which is the smallest detectable change in pressure.

A 12-bit A/D converter provides 12 significant bits in the digital value and has a resolution of 1 part in 4096 = 2^{12}. The input voltage is converted to one of 2^{12} = 4096 states, which computer people always number 0, 1, 2, ... , 4095. For industrial applications, the input voltage range may be either of the following:

- The input voltage range is 0 to 5.00 V. An input of 5.00 V corresponds to 4095, and the resolution is 1 part in 4096.
- Computer people just can't resist powers of 2! Even though often referred to as a 5-V A/D, the input voltage range is 0 to 5.12 V ($512 = 2^9$). Thus 5.12 V is converted to 4095 counts. An input of 5 V gives a count of 4000, so the resolution over 0 to 5 V is 1 part in 4000. The effective resolution is not quite 12 bits ($\log_2 4000 = 11.97$), but it is usually said to be 12 bits.

As in the previous example for the input from the pressure transmitter, we shall generally use a resolution of 1 part in 4000 for a 12-bit A/D. The reason is simple—the numbers are easier to work with. For the pressure transmitter, 800 mm Hg/4000 counts = 0.2 mm Hg/count, but 800 mm Hg/4096 counts = 0.1953125 mm Hg/count.

The A/D converters used in industrial systems have resolutions between 11 bits (1 part in 2000) and 15 bits (1 part in 32,000).

Current Inputs   In industrial systems, most analog inputs are via a circuit known as a *current loop*. We will examine this in more detail later and explain the reasons for using DC current instead of DC voltage. The input range for the current signal is 4 to 20 ma. However, the A/D converter accepts volts, not current. By inserting a 250-Ω resistor (known as the *range resistor*) into the circuit, a current of 4 ma is converted to 1 V and a current of 20 ma is converted to 5 V.

Suppose a pressure measurement device generates 4 ma when the pressure is 0 mm Hg absolute and 20 ma when the pressure is 800 mm Hg absolute. An A/D converter with a resolution of 1 part in 4000 would generate the following digital values:

Quantity	Lower Range Value	Upper Range Value
Absolute pressure (mm Hg)	0	800
Current (ma)	4	20
Input (V DC)	1	5
Raw value (count)	800	4000

The effective input range is not 0 to 5 V; it is really 1 to 5 V. This decreases the effective resolution from 1 part in 4000 to 1 part in 3200. In terms of the pressure, the resolution is 800 mm Hg/3200 counts = 0.25 mm Hg/count, which is now the smallest detectable change in pressure.

Even though the A/D converter has a 12-bit resolution, the effective resolution of the input system is no longer 12 bits or 1 part in 4000. Instead, the effective resolution is 1 part in 3200. Sometimes this is stated as a number of bits, but with a fractional part. For this example, the resolution for the input value could be stated as $\log_2 3200 = 11.6$ bits.

At the expense of a small increase in complexity, some input systems convert the 4- to 20-ma signal into a 0- to 5-V signal so that the entire input range of the A/D can be used. Basically, one has to pay close attention to the details when examining the resolution of an input value.

Engineering Value Inputs   With the incorporation of digital technology into the I/O equipment, providing a raw value in engineering units is becoming a common practice. One example is an resistance temperature detector (RTD) input card that performs the conversion to temperature units. Similar cards are available for thermocouples.

The *raw value* is the temperature in °C (or °F) with a resolution of 0.1 °C (or 0.1 °F). The raw value is represented as an integer number, so fractional parts are never present in the raw value. The location of the decimal point is at a fixed position, specifically, one decimal digit from the right. Therefore, a raw value of 1097 means a temperature of 109.7 °C (or 109.7 °F). Negative temperatures give negative raw values.

The limits on the measured value are not imposed by the I/O system; they are imposed by the RTD, specifically, −200° to 850 °C for the Class B RTDs normally installed in industrial processes. A resolution of 0.1 °C is a resolution of 1 part in 10,500 for the entire possible measurement range. But when measuring temperatures in water-based processes at atmospheric pressure (such as fermentation processes), the range of interest is 0° to 100 °C. A resolution of 0.1 °C becomes a resolution of 1 part in 1000 for the range of interest.

Display Resolution Vs. Internal Resolution   When determining the resolution for a measured value, the tendency is to consider the display resolution from the perspective of what some human needs to see, such as, How many digits after the decimal point does the process operator need to see? or How many digits after the decimal point are required in the historical data records? This is the *precision* as the term is used in computer circles.

But the resolution is also important for whatever processing (such as control calculations) is being performed within the system. Suppose a process pressure is being measured over the range 0 to 10 psig. For values on the operator displays, a resolution of 0.1 psig is probably adequate. This is 1 part in 100, or 1% of the measurement range. Now suppose this pressure measurement is the input to a pressure controller with a gain of 5%/%. Each time the pressure changes by 0.1 psig (1% of the measurement range), the controller output will abruptly change by $1\% \times 5\%/\% = 5\%$. The bump to the process will be noticeable.

This issue is especially important for I/O systems that provide engineering value inputs. The resolution provided in the input value becomes the system's internal resolution. This resolution must be adequate for whatever internal processing, such as control calculations, is being performed by the system. If the internal resolution is more than required, the resolution can be reduced when the data are output to displays or historical files.

1.5. ZERO, SPAN, AND RANGE

As Figure 1.9 illustrates, the output of most modern transmitters varies linearly with the value of the measured variable or process variable. Such transmitters require only two adjustments:

Zero. Determines the value of the measured variable for which the transmitter output is equal to its lower-range value.

Span. Determines the change in the measured variable required to cause the transmitter output to change from is lower-range value to its upper-range value.

The zero and span adjustments determine the measurement range for the measured variable. The range can also be determined from the following two values, both in the engineering units of the measured variable:

Lower-range value. Equal to the zero adjustment.
Upper-range value. Equal to the zero adjustment plus the span adjustment.

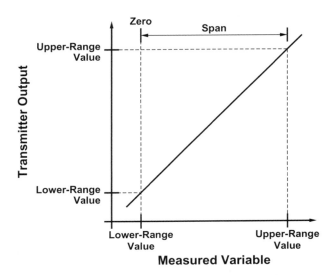

Figure 1.9. Operating line for a linear transmitter.

Specifying Zero and Span

Measurement devices based on the older technology contained physical adjustments for setting the zero and the span. The zero adjustment had to be set first, and then the span adjustment could be set.

But with microprocessors incorporated into measurement devices, the physical adjustments are replaced by software coefficients for the zero and span. Values for these coefficients can be obtained as follows:

- Specify the zero and span explicitly.
- Specify the zero and one point on the measurement device's operating line. The span is calculated from this point on the operating line.
- Specify one point on the measurement device's operating line and the span. The zero is calculated from this point on the operating line.
- Specify two points on the measurement device's operating line. The zero and span are calculated from these two points on the operating line.

Overrange

As the graph in Figure 1.9 suggests, many transmitters will function somewhat outside their measurement range. Such transmitters provide a small overrange. The amount of overrange varies, but typically is in the vicinity of 5% of the span. However, it is not recommended that the transmitter be routinely used in the measurement region provided by this overrange.

Some systems provide range alarms to provide an alert when the input from the measurement device is beyond the measurement range. These alarms usually include a deadband, a typical value being 0.5% of the span. If the lower-range value for the transmitter output is considered to be 0% and the upper-range value for the transmitter output is considered to be 100%, the typical logic for the range alarms is expressed as follows:

Occurrence. The range alarm is issued when the input from the measurement device is less than –0.5% or greater than 100.5%.

Return to normal. The range alarm returns to normal when the input from the measurement device is greater than or equal to 0% and less than or equal to 100%.

Zero Error

Figure 1.10 illustrates the behavior of a linear transmitter with an error in the specification for the zero. As shown, the transmitter output is above its lower-range value (typically 4 ma) when the variable being sensed is equal to its lower range value.

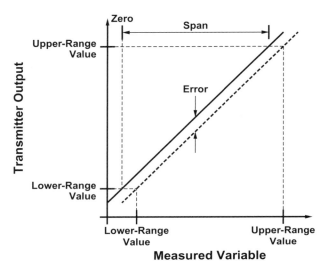

Figure 1.10. Zero error.

If the span specification is exact, any zero error results in a constant error over the entire measurement range. *Zero drift* is the change in the zero adjustment with time.

Zero drift normally causes the zero error to increase. One of the advantages of microprocessor-based (smart) measurement devices is that the zero error and zero drift are smaller than in conventional transmitters.

In some applications, a small zero error can have significant consequences. Consider a flow transmitter that provides the input to a flow controller. When a value of zero is specified for the set point, the flow controller is expected to completely close the valve. Whether this occurs or not depends on the nature of the zero error. There are two possibilities:

- The transmitter indicates zero flow when the flow is slightly positive. As the flow controller responds to the output of the flow transmitter, the flow controller will attempt to position the control valve to a very small opening to achieve the flow required for the transmitter output to be zero. This behavior is unacceptable.
- When the flow is zero, the transmitter output is a small value, suggesting a small flow. In attempting to drive the measured value of the flow to zero, the flow controller will decrease its output to the control valve as much as possible (to a value known as the lower output limit). This is the desired behavior.

As the latter is the preferred behavior, the zero for the flow measurement is often intentionally biased so that it will indicate zero for a slightly positive flow.

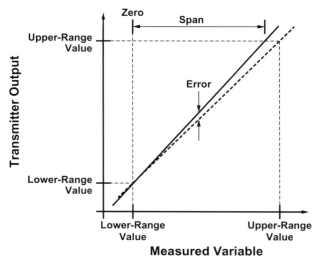

Figure 1.11. Span error.

Span Error

Figure 1.11 illustrates the behavior of a linear transmitter with an error in the specification for the span. In this example, a change in the measured variable from its lower-range value to its upper-range value gives a change in the transmitter output of less than its nominal span (16 ma for current loop installations).

If the zero specification is exact, the error is zero when the measured variable is equal to the lower-range value of the measurement range. The error increases as the process variable approaches the upper-range value of the measurement range.

Nonlinear Transmitter

Figure 1.12 illustrates the behavior of a transmitter whose output exhibits a modest degree of departure from linear behavior. If the zero and span specifications are exact, the error is zero when the measured variable is equal to either the lower-range value or the upper-range value of the measurement range.

The error in such a transmitter depends on the nature of the nonlinearity. This is often summarized as a statement for the maximum departure from linear behavior.

If the nature of the nonlinearity is known, digital systems permit a characterization function (or its equivalent) to be applied to its input value (which

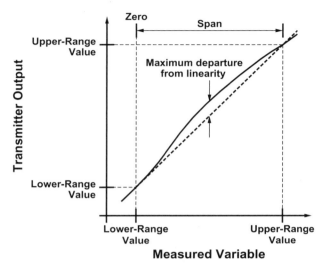

Figure 1.12. Nonlinear transmitter.

is the output of the transmitter) to compensate the measurement for its nonlinear behavior. For microprocessor-based measurement devices, the manufacturers normally incorporate such characterization functions into the measurement device itself, which usually reduces the maximum departure from linearity to an acceptable value.

Gain or Sensitivity

The gain or sensitivity of a device quantifies the degree to which a change in the input will affect the output of the device. For a measurement device, the input is considered to be the variable being measured, expressed in engineering units. The output is considered to be the output signal from the measurement device.

For a linear measurement device (including those with an insignificant departure from linearity), its gain or sensitivity (usually called the transmitter gain) is

$$\text{Transmitter gain} = \frac{\text{Span of the output signal}}{\text{Engineering span of the measured variable}}$$

The following calculations give the gain of a temperature transmitter with a measurement range of $50°$ to $250°F$:

Signal Type	Signal Span	Gain or Sensitivity
Current loop	4–20 ma	$\dfrac{16\,\text{ma}}{200\,^{\circ}\text{F}} = 0.08\,\text{ma}/^{\circ}\text{F}$
Digital	0–100%	$\dfrac{100\%}{200\,^{\circ}\text{F}} = 0.5\%/^{\circ}\text{F}$

For electronic systems, the signal is the current loop with an output range of 4 to 20 ma. For digital systems, the output range is normally considered to be 0 to 100%. A gain or sensitivity of 0.5%/°F means that the transmitter output changes by 0.5% for each 1 °F change in temperature.

1.6. TURNDOWN RATIO AND RANGEABILITY

Although the turndown ratio can be applied to other situations, it is usually best understood for anything to which the term *throughput* is meaningful. Let's explain it in the context of a process such as a steam boiler.

Suppose the maximum capacity of a boiler is 50,000 lb/hr of steam. Does this mean that it is possible to operate the process at any steam rate below 50,000 lb/hr? Definitely not. In addition to the maximum steam rate, the designers of the boiler are given a specification for the minimum steam rate at which the boiler is to be operated. Alternatively, the designers can be given a specification for the boiler's turndown ratio, which is defined as follows:

$$\text{Turndown ratio} = \frac{\text{Maximum steam capacity for the boiler}}{\text{Minimum steam rate at which the boiler can be operated}}$$

If the boiler can be operated down to 20,000 lb/hr, its turndown ratio is 2.5:1.

Flow Measurements

In some applications, the flow is large at times but small at other times. How well a flow measurement device will perform in such applications depends on its turndown ratio, which is defined as follows:

$$\text{Turndown ratio} = \frac{\text{Maximum measurable flow}}{\text{Minimum nonzero measurable flow}}$$

The maximum measurable flow is usually taken to be the upper-range value for the measurement range. The lower-range value for the measurement range is usually zero. But we have to be careful with zero. Some flow meters, specifically, the vortex shedding meter, will indicate zero even if there is a small flow through the meter. The interest is the smallest nonzero flow that the meter can read in accordance with its specifications. The minimum measurable flow could be considered to be the smallest flow for which the flow measurement device can be calibrated.

One limitation of an orifice meter is that its turndown ratio is rather poor; typical values are in the range of 3 : 1 to perhaps 5 : 1. Other types of flow meters usually perform much better. For example, 50 : 1 is a typical value for the turndown ratio of a magnetic flow meter. If the capacity of such a meter is 100 gal/min, it can also accurately measure a flow of 2 gal/min.

Other Measurements

For a given measurement device, the maximum possible measurable value is the upper-range value. The minimum possible measurable value is the lower-range value. The maximum possible turndown ratio is as follows:

$$\text{Turndown ratio} = \frac{\text{Upper-range value}}{\text{Lower-range value}}$$

Obviously this equation has problems when the lower-range value is zero, which is unfortunately the case for most measurements for which the turndown ratio is of interest. We have already discussed flow measurements. Turndown ratio is often applied to pressure measurements, where the lower-range value is often zero (either zero absolute or zero gauge). Instead of using the lower-range value to compute the turndown ratio, the minimum nonzero measurable pressure must be used.

In industrial applications, turndown ratio is rarely of interest for temperature measurements, composition measurements, or measurements of other intensive properties.

Rangeability

The turndown ratio pertains to the performance of a measurement device with the existing settings for the zero and span. Rangeability basically assesses the ability of a given measurement device to be applied to a variety of applications. Rangeability is defined as follows:

$$\text{Rangeability} = \frac{\text{Maximum possible upper-range value}}{\text{Minimum possible upper-range value}}$$

Most measurement devices permit the upper-range value to be changed so that a given device can be applied to a range of applications. For flow measurements, some applications will require large flows; other applications will require much smaller flows. For example, if a flow measurement device can have a measurement range as low as 0 to 20 lpm but as high as 0 to 200 lpm, its rangeability is 10:1.

A high rangeability may not necessarily require a high turndown ratio. In one application, the flow may vary from 15 to 20 lpm, but in another application, the flow may vary from 150 to 200 lpm. In each application, a turndown ratio of 2:1 is adequate.

1.7. ACCURACY

Accuracy is generally thought of as the conformity of the measured value with the true value. However, how do we know the true value? We never know the true value.

To establish the accuracy of a measurement device, we compare the indicated value (the *measurand*) with that of a standard used for calibration or with a device whose accuracy is considered to be far better than that of our measurement device. An example is the use of a dead weight tester to determine the accuracy of a pressure measurement device.

Consequently, accuracy is the degree of conformity of the measured value with either a standard, reference, or other accepted value for the variable being measured. Accuracy is usually stated as the error in the measured value. Therefore, it can be argued that it is actually the inaccuracy of the measurement device that is being stated. However, the practice in the industry is to refer to these as stating the accuracy of the measurement device.

Bases for Stating Accuracy

The basis on which the accuracy of a measurement device is stated depends on the principles that are employed to design the measurement device. For a temperature transmitter with a measurement range of 50° to 250°F, the four possibilities are of as follows if the current reading is 150°F:

Basis	Example
Percent of span	$\pm 1\%$ of 200°F = ± 2°F
Percent of reading	$\pm 1\%$ of 150°F = ± 1.5°F
Percent of upper-range value	$\pm 1\%$ of 250°F = ± 2.5°F
Absolute accuracy	± 1°F

Process operations normally depend on the absolute accuracy stated as a value in the engineering units of the measured variable. To compare performance, all expressions of accuracy must be translated to this basis. However, if the configuration parameters of the measurement device are altered, the absolute accuracy may change.

Traditionally, accuracy was most frequently stated as percent of span. But with microprocessor-based measurement devices, percent of upper-range value has become very common. In some cases, these are equivalent. For example, the measurement range of most flow measurement devices is from zero to an upper-range value. In such cases, percent of span and percent of upper-range value are the same.

But consider a microprocessor-based temperature transmitter whose measurement range is $0°$ to $600°F$. If the accuracy is stated as percent of upper-range value, changing the measurement range to $500°$ to $600°F$ has no effect on the absolute accuracy (in $°F$). But if the accuracy is stated as percent of span, changing the measurement range to $500°$ to $600°F$ improves the absolute accuracy by a factor of six.

Accuracy as Percent of Reading

For some applications, accuracy stated as percent of reading is preferred over accuracy expressed as either percent of span or percent of upper-range value. A flow meter whose accuracy is stated as percent of reading performs far better at low flows than does a flow transmitter whose accuracy is stated as percent of upper-range value. The error expressed as a percent of reading increases rapidly as the flow decreases. The relationship is as follows:

$$\frac{\text{Accuracy as percent}}{\text{of reading}} = \frac{(\text{Accuracy as percent of upper-range value}) \times F_{\text{MAX}}}{F}$$

where F_{MAX} = upper-range value for the flow measurement, F = measured value for flow.

For a measurement device with a stated accuracy of 1% of upper-range value, Figure 1.13 presents the accuracy as a percent of reading as a function of the measured value of the flow. On the basis for percent of reading, the accuracy is very poor at low flows. This behavior often limits the minimum flow for which the measured value is acceptable.

Limitations

Suppose a manufacturer states the accuracy of a viscosity measurement as "±2% for Newtonian fluids of low viscosity." In industrial practice, the viscosity of a Newtonian fluid of low viscosity seems rarely to be of much interest. But from the manufacturer's perspective, non-Newtonian behavior can be in a variety of forms. Basically, it becomes the responsibility of the user to translate this statement into something that applies to a specific application. The potential of a sale creates an incentive on the part of the manufacturer to assist the user in making this translation. However, the user will have to provide information on the behavior of the fluid whose viscosity is to be measured.

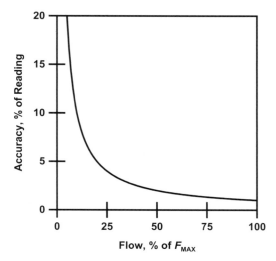

Figure 1.13. Absolute accuracy as a function of flow for a meter with a stated accuracy of 1% of the upper-range value.

Many statements of accuracy are accompanied by statements regarding process conditions. Furthermore, the accuracy is sometimes stated for measuring some property of air or water, often at or near ambient conditions. Conditions within the process can be far removed from such conditions.

Installation

A measurement device is not always directly exposed to the process variable whose value is of interest. For example, the temperature probe in Figure 1.14 is installed within a thermowell to measure the fluid temperature T_F. The accuracy statements from the supplier of the temperature probe will apply to the probe temperature T_P, not to the fluid temperature. Often we casually assume that the probe temperature is the same as the fluid temperature, but is this really correct?

Manufacturers often advise as to proper installation procedures, but it is ultimately the user's responsibility to install a measurement device properly. For temperature probes inserted into thermowells, spring-loaded assemblies apply pressure to ensure good metal-to-metal contact between the tip of the probe and the thermowell. Alternatively, the thermowell can be filled with a heat-conducting liquid.

Users are also advised to insert the thermowell at a location where the fluid motion is significant and to keep the mass of the probe as low as possible. But even if all such advice were followed, would the temperature of the probe be exactly the same as the fluid temperature? Probably not. At process operating conditions, how different can be difficult to ascertain.

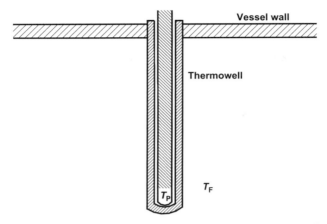

Figure 1.14. Temperature probe inserted into a thermowell.

1.8. REPEATABILITY

Although somewhat controversial, the contributions to the error in a measured value have traditionally been viewed as of two types:

Random. Errors of this type lead to the scatter observed when repeated measurements are made. This type of error can be quantified by computing the standard deviation or variance for a set of measurements.

Systematic. Errors of this type lead to a bias or offset between the measured value and the true value (which in practice is never known). Repeated measurements using the same device will not detect systematic errors. Even a set of measurements with a small standard deviation could be significantly different from the true value, whatever that is.

Quantifying the systematic error is far more difficult than quantifying the random error. Calibration attempts to address this issue by comparing the result to the value indicated by a more accurate measurement device. However, the possibility exists that there may be bias in the value from the more accurate device and even that this bias is the same as the bias in the device being calibrated. Comparing devices that rely on different measurement principles (for example, orifice meter vs. turbine meter) reduces the probability that the two biases will be the same, but not necessarily to zero.

Concept

Repeatability focuses only on the random errors and basically ignores systematic errors or biases. It applies to a given measurement device in a given application and can be viewed from two perspectives:

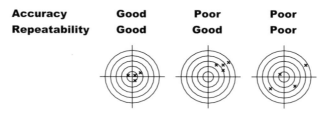

Figure 1.15. Accuracy and repeatability.

- The process conditions are exactly the same on two different occasions. The difference in the measured value on these two occasions reflects the repeatability (or actually the nonrepeatability) of the measurement device.
- The measured values are exactly the same on two different occasions. The difference in the process conditions on these two occasions reflects the repeatability (or actually the nonrepeatability) of the measurement device.

The scatter patterns for the targets in Figure 1.15 illustrate the concepts of accuracy and repeatability.

Repeatability is similar to accuracy in that it can be expressed on any of the following bases:

- Percent of span.
- Percent of reading.
- Percent of upper-range value.
- Absolute repeatability (in engineering units).

Accuracy Vs. Repeatability

For some measurement applications, accuracy is crucial, but for others, repeatability is crucial. Figure 1.7 illustrates an off-line analysis in which a sample is taken to the quality control (QC) lab for analysis, and the result's reported to the process operator. Suppose the operator adjusts the target for a temperature controller based on the results of the analysis. In this application:

QC lab analytical instrument. Good accuracy is required. Bias in this measured value must be minimized.

Process temperature measurement. Good repeatability is required. Bias in this measured value is not a problem. Often the operator is thinking in terms of changes; to make a certain change in the results of the analysis, the operator knows by experience what change is required in the process temperature. Modest errors in the measured value of the process temperature do not impair the operator's performance, provided the errors are the same from day to day.

The requirements for accuracy vs. repeatability are often summarized as follows:

- Regulatory control is focused on maintaining constant conditions within the process. This can be achieved provided the measurement device has good repeatability.
- Endeavors such as process optimization and constraint control depend on the accuracy of the measurement device.

Reproducibility

Repeatability applies to a given measurement device in a given application. *Reproducibility* applies to different measurement devices in the same application. There are two scenarios:

- A given measurement device fails and is replaced "in kind"—that is, maintenance removes the faulty device and installs a new one of the same model. The agreement (or actually the nonagreement) of the measured values from the two devices reflects the reproducibility for this specific model of the measurement device.
- A given measurement device is replaced by a measurement device based on a different measurement principle. For example, an orifice meter is replaced by a mass flow meter. How do the measured values from the mass flow meter agree with the measured values from the orifice meter? Rarely are the differences trivial. If you abruptly switch from basing process operations on the orifice meter to basing process operations on the mass flow meter, the consequences are likely to be noticeable. After installing the mass flow meter, you should continue to run the process based on the measured values from the orifice meter. Note the difference (the nonreproducibility) between the measured values from the two meters and establish targets for the mass flow meter that reflect the current manner of process operation. Using these targets will give a smooth transition from one meter to the other. Thereafter, the targets can be adjusted as warranted.

Precision

When reading specifications for measurement devices, you will encounter the term *precision*. But as used, this term unfortunately has a variety of definitions, in fact, so many that the National Institute for Standards and Technology (NIST) does not use it. But at least there is agreement on one aspect:

Precision is not another word for accuracy.

In some cases, precision is used as another term for repeatability. When a set of measured values is available, precision is sometimes considered to be the standard deviation of the measured values. Occasionally, precision is used as another term for resolution—that is, the number of significant figures in the measured value. Sometimes the manner in which precision is used will be clear, but too often it is not. Make no assumptions here. If you are relying on this term, have the supplier provide its definition and make sure you understand it.

Measurement Uncertainty

Suppose a temperature measurement device with a manufacturer-stated accuracy of ±0.5 °C indicates 152.1 °C. This is generally understood to mean that the true value is between 151.6 °C and 152.6 °C. How did the manufacturer conclude that the accuracy is ±0.5 °C? You will have to ask the manufacturer. There is no generally agreed on method for establishing assessments of the accuracy of measurement devices.

Does 152.1 °C with an accuracy of ±0.5 °C imply that the true value is more likely to be 152.1 °C than 151.6 °C or 152.6 °C? Except when there are specific reasons to believe otherwise, a normal or Gaussian distribution is usually assumed. But unless the manufacturer can provide additional information, no conclusions on the probability of values within the interval of 151.6 °C to 152.6 °C can be definitively made. Any probability distribution function, including those in Figure 1.16, is possible.

The current direction is to use the term *measurement uncertainty* in lieu of accuracy, repeatability, and the like to quantify the performance of a measurement device. Measurement uncertainty quantifies the probability associated with a measured value. NIST now views accuracy, repeatability, and so on as

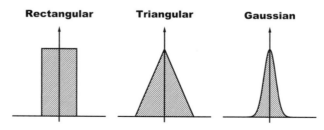

Figure 1.16. Possible distribution functions for a measured value.

qualitative terms. However, most manufacturers of measurement devices continue to state quantitative values for these.

1.9. MEASUREMENT UNCERTAINTY

To understand the need for developing an expression for the uncertainty associated with a measurement, the following statement must be recognized:

It is impossible to obtain the true value for any measurand.

Measurement is not an exact science. Therefore, it is imperative to agree on:

- How to express the uncertainty associated with a measured value.
- The approach used to obtain quantitative values for the uncertainty associated with a measured value.

The *International Vocabulary of Basic and General Terms in Metrology*[1] (VIM) defines *uncertainty* as a parameter associated with the result of a measurement that characterizes the dispersion of the values that could be reasonably attributed to the measurand.

The *Guide to the Expression of Uncertainty in Measurement*[2] (GUM) is the controlling document. However, it is generic to all types of measurement devices; documents specific to a given type of measurement device provide the details. For example, ISO 5167-2 *Measurement of Fluid Flow by Means of Pressure Differential Devices Inserted in Circular Cross-Section Conduits Running Full—Part 2: Orifice Plates*[3] provides a method for calculating the measurement uncertainty for an orifice meter.

Statistical Approach

When multiple measurements are available, statistical methods can be applied to obtain a value for the measurement uncertainty. Consider the following example of multiple measurements of a temperature:

Number of measurements	10
Measured values	152.2 °C, 151.7 °C, 152.1 °C, 152.2 °C, 152.0 °C, 152.1 °C, 152.0 °C, 152.3 °C, 152.2 °C, 152.2 °C
Mean value	152.1 °C
Standard deviation	0.17 °C
Degrees of freedom	9
Coverage factor	2.32 (95% confidence level)
Uncertainty	0.39 °C (95% confidence level)

The measurement uncertainty is computed as follows:

1. Compute the mean (152.1 °C) and standard deviation ($\sigma = 0.17$ °C).
2. With 10 measurements, there are 9 degrees of freedom. From the Student's t-table, the coverage factor k for a 95% level of confidence is 2.32 (actually, this is for 95.45%, which corresponds to 2σ for a normal distribution).
3. The measurement uncertainty is $k\sigma = 0.39$ °C.

The measured value would be stated as 152.1 °C with a measurement uncertainty of 0.39 °C at a 95% level of confidence.

Continuous Measurements

There are a couple of problems with the statistical approach:

- For continuous measurement devices, multiple measurements are not available.
- The GUM does not view this approach favorably, mainly because no effort is expended to understand and analyze the measurement process.

Shortly we will discuss issues, such as false accepts and false rejects, associated with decisions made on measurements of *critical operating variables*. When a measured value is from multiple measurements made on one or more samples, the statistical approach can be used to establish the measurement uncertainty.

But what if one or more of the critical operating variables are continuous measurements? The most convenient approach would be to determine a measurement uncertainty for the continuous measurements that is equivalent to the measurement uncertainty established by the statistical approach. The measurement decision procedures would then be independent of whether the measurement is sampled or continuous.

Unfortunately, this proves to be a challenge. An approach will be described next, but the discussion also includes at least some of the criticisms of the approach.

The GUM Approach

The GUM provides a framework for evaluating measurement uncertainty based on a fundamental understanding of the measurement process. The following six steps summarize this approach.

1. Specify the measurand.
2. Model the measurement.

3. Quantify the contributors to uncertainty.
4. Determine the sensitivity coefficient for each contributor.
5. Combine the contributors.
6. Calculate the expanded uncertainty.

We will provide only an overview of this approach as it is applied to industrial measurement devices.

Critics have correctly observed that the resulting value for the measurement uncertainty is a calculated value that is not confirmed by subsequent testing and experimentation. One is advised to apply "reasonableness tests" to the final result and to be suspicious of very small and very large values for the uncertainty. One can even raise the issue of the uncertainty associated with the value obtained for the measurement uncertainty. Evaluating the measurement uncertainty is clearly not an exact process either.

The process should never be considered an exercise merely to get a number for the measurement uncertainty. The analysis quantifies the influence of each contributor to the overall measurement uncertainty. Knowing the influence of each contributor permits the technology pertaining to the measurement device to be enhanced. Despite its shortcomings, the GUM approach is superior to the alternatives and is clearly the direction of the future.

Measurement Model   The analysis of a measurement device is based on a mathematical model that relates the measurement result y to the various inputs x_j that affect the measurement result:

$$y = f(x_1, x_2, \ldots, x_n)$$

where y = measurement result; x_j = input that affects measurement result; n = number of inputs that affect measurement result.

Clearly this model is specific to each type of measurement device. For example, ISO 5167-2[3] provides such an equation for the orifice meter.

Each input x_j is potentially a contributor to the measurement uncertainty. Any variability in an input will lead to variability in the measurement result. For an orifice plate, ISO 5167-2[3] specifies that the orifice must be circular to a certain tolerance, specifically, "no diameter shall differ by more than 0.05% from the value of the mean diameter." This raises two issues:

- The uncertainty for the orifice diameter u_d must be quantified. The tolerances in the standard normally provide the basis for quantifying this uncertainty, but it is possible that a supplier will use tighter tolerances in the manufacturing process.
- The contribution of the uncertainty for the orifice diameter must be translated to uncertainty for the measured value of the flow. ISO 5167-2[3] provides an equation relating the flow to various quantities, one

of which is the orifice diameter. The equation permits the quantitative translation of uncertainty in the orifice diameter to uncertainty in the measured flow.

Sensitivity Coefficients   The sensitivity coefficient c_j translates a small change in input x_j to a change in the measurement result y:

$$c_j = \frac{\partial y}{\partial x_j} \cong \frac{\Delta y}{\Delta x_j}$$

where c_j = sensitivity coefficient for input j.

A sensitivity coefficient can be thought of as the change in the measured value y that would result from a one-unit change in input x_j. The engineering units for the sensitivity coefficient are the engineering units for the measured value over the engineering units for the input.

The sensitivity coefficients can be evaluated either

Analytically. This involves taking the partial derivative of the measurement model. While preferred, this is not always possible, and even when possible, the resulting expression could be excessively complex.

Numerically. This is based on the finite difference approximation to the partial derivative. To perform the calculation, values must be provided for all inputs. Most measurement models are nonlinear equations, so the values provided for the inputs will affect the value calculated for the sensitivity coefficient.

Combined Standard Uncertainty   For input x_j, the contribution to the uncertainty in the measured value is the product of the uncertainty u_j for that input and the sensitivity coefficient c_j for that input. The combined effect of all inputs gives the combined standard uncertainty u_c for the measured value. There are two possibilities:

Inputs are uncorrelated. The individual contributions are combined using the root mean square equation:

$$u_c = \left[\sum_{j=1}^{n} c_i^2 u_i^2 \right]^{1/2}$$

where u_c = combined standard uncertainty for measured value; u_j = uncertainty for input j.

Inputs are Correlated. Although we will not present it, an equation is available to compute the combined standard uncertainty for situations in which some degree of correlation exists between the inputs. The equation is more complex, but computationally presents no real problems. However, this equation contains another term, specifically the correlation coefficient $r(x_j, x_k)$, which characterizes the degree of correlation between input j and input k. Because we rarely have a good source for this term, it is usually estimated from the standard deviations of the two inputs.

It is tempting to make the assumption that the inputs are uncorrelated. But to the extent that this assumption is not correct, the value computed for the combined standard uncertainty will be low.

Expanded Uncertainty Measurement uncertainty is normally stated as $y \pm U$, where U is the expanded uncertainty. This interval encompasses the possible values for the measurand with a given level of confidence (typically 95%). To obtain the expanded uncertainty U, the combined standard uncertainty u_c is multiplied by the coverage factor k:

$$U = ku_c$$

where U = expanded undertainty; k = coverage factor.
   The coverage factor k is obtained from the Student's t-table (extensive versions are provided in numerous reference books). To use this table requires values for the following:

Level of confidence. This is typically 95% (or 94.45% for 2σ), although 99.7% (for 3σ) is occasionally used.

Degrees of freedom v_y *for the measurand.* This is computed from the degrees of freedom v_j for each input using the Welch-Satterthwaite equation:

$$v_y = \frac{u_c^4}{\displaystyle\sum_{j=1}^{n} \frac{c_i^4 u_i^4}{v_j}}$$

For measurement devices, obtaining a value for the degrees of freedom v_j for each input x_j is a major problem.

For a 95% confidence level, the minimum value of the coverage factor k is 2 (infinite degrees of freedom). As the number of degrees of freedom decreases, the coverage factor increases, but it is unlikely that k would exceed 3.

Degrees of Freedom When the uncertainty for an input is determined from a set of measurements, the degrees of freedom are $n - 1$, or one less than the number of measurements in the set. But for the orifice meter example used previously, the uncertainty for the orifice diameter is determined from specifications in the standard. How do we determine the degrees of freedom for this input?

One possibility is to assume that the degrees of freedom are infinite. If this is done for all inputs, the coverage factor k for a 95% confidence level would be 2, which is the lowest possible value. Making such an assumption would give the smallest possible value for the expanded uncertainty U.

The value for the uncertainty u_j for input j is not exact, which should be no surprise. Let Δu_j be the uncertainty for the uncertainty u_j. The relative uncertainty is $\Delta u_j / u_j$. The GUM provides the following equation for computing the degrees of freedom v_j for input j from the relative uncertainty for input j:

$$v_j = \frac{0.5}{\left[\Delta u_j / u_j\right]^2}$$

But where do we get a value for the relative uncertainty? Unfortunately, no method other than professional judgment has been proposed for obtaining a value for the relative uncertainty $\Delta u_j / u_j$.

Current Status

I hope you now have a general idea of how to determine the measurement uncertainty for an industrial measurement device. Could you do it? Not based on only what has been presented in this rather superficial treatment of the subject. The evaluation of measurement uncertainty is a subject for an entire book.

In areas where a measurement test can be repeated to give multiple samples, a value for the measurement uncertainty now routinely accompanies test results. An informative source of information is the *A2LA Guide for the Estimation of Measurement Uncertainty in Testing*.[4]

For industrial measurement devices, progress has been much slower; most of the push is coming from the European community. The application of measurement uncertainty to industrial measurement devices is difficult, and the critics have adequately (and generally correctly) noted the problems. At this time, the manufacturers routinely state values for the traditional measures of accuracy, repeatability, etc., but not for measurement uncertainty. In the subsequent topics on specific types of measurement devices, this forces us to use the older measures in lieu of measurement uncertainty. This will change; the question is not *if*, but *when*. It is clear that measurement uncertainty is the direction of the future.

1.10. MEASUREMENT DECISION RISK

A process is operating properly if every critical operating variable is within a specified tolerance of a target. A product is acceptable if every product quality measure is within the tolerances stated in the specifications.

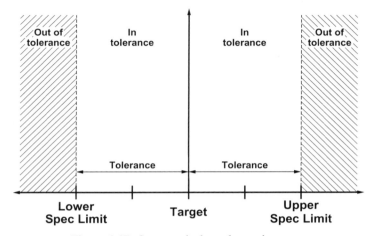

Figure 1.17. Symmetrical product tolerances.

The possibilities for process/product tolerances include the following:

Symmetrical. The same tolerance is applied to each side of the target. For example, a temperature should be 150 °C ± 10 °C. Temperatures above 160 °C or below 140 °C are unacceptable. A symmetrical tolerance is illustrated in Figure 1.17.

Nonsymmetrical. Different tolerances are applied to the two sides of the target. For example, a temperature should be 145 °C + 15 °C/−5 °C. Temperatures above 160 °C or below 140 °C are unacceptable.

One sided. A tolerance is applied to only one side of the target. For example, a temperature should be 145 °C + anything/−5 °C. Any temperature above 140 °C is acceptable.

All of these cases can be analyzed. However, we will consider only the symmetrical case.

Process/Product Distribution

Although the objective is to maintain the process or product exactly at the target, there will always be excursions from the target. The magnitude of the excursions depend on

- Magnitude of the upsets to the process.
- Performance of the process operators, control systems, etc. responsible for maintaining the process at the target.

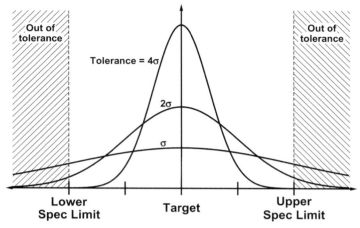

Figure 1.18. Distribution functions for tolerance equal to various multiples of the standard deviation.

The assumption is usually made that the variations are Gaussian in nature. If so, they can be characterized by their standard deviation σ. Figure 1.18 presents the distribution functions for three cases:

Tolerance equals twice the standard deviation σ. The out-of-tolerance is 4.6%. That is, the process is operating outside the acceptable limits approximately 4.6% of the time, or about 4.6% of the product is not within specifications.

Tolerance equals the standard deviation σ. The out-of-tolerance is 31.8%, which for most production operations is unacceptably large.

Tolerance equals four times the standard deviation σ. The out-of-tolerance is essentially zero.

Figure 1.18 applies to either of the following situations:

Constant standard deviation σ*; variable tolerance.* The process variations remain the same, and the product specifications are changed. The product specifications invariably become tighter, which means an increase in the out of tolerance for constant process variations (same process operations).

Constant tolerance; variable standard deviation σ. The product specifications remain the same, and the process variations are reduced. This case applies when automation technology is applied to provide more uniform process operations to reduce the out of tolerance.

The customary scenario is that the specifications become tighter, which provides the incentive to enhance the process and/or the controls to maintain the out of tolerance at an acceptable level.

Measurement Distribution

Figure 1.19 presents a Gaussian distribution function of the measurement results when the process is exactly at the target. This is commonly assumed unless there are specific reasons to believe otherwise.

If the process is indeed at the target, the measurement result is not necessarily equal to the target. But for the distribution function illustrated in Figure 1.19, the probability that the measurement result could be outside the tolerance limits for the process is extremely small. But when the process is close to one of the specification limits, there are four possibilities:

Process within limits; measurement result within limits. Process correctly considered to be in tolerance.

Process within limits; measurement result outside limits. Process incorrectly considered to be out of tolerance. For products, this is a *false reject.*

Process outside limits; measurement result within limits. Process incorrectly considered to be in tolerance. For products, this is a *false accept.*

Process outside limits; measurement result outside limits. Process correctly considered to be out of tolerance.

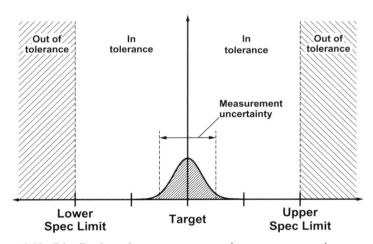

Figure 1.19. Distribution of measurement results, process operating at target.

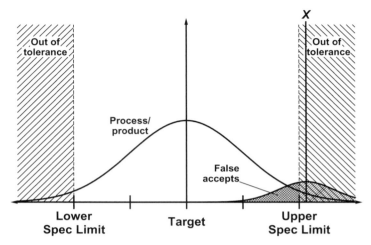

Figure 1.20. False accepts.

False Accepts   Figure 1.20 depicts a situation in which the true value is slightly outside the upper specification limit. The process is actually out of tolerance. But because of the measurement uncertainty, the measurement result could be on the acceptable side of the upper specification limit. For products, this leads to what is referred to as false accepts. In industries like pharmaceuticals, false accepts are understandably a serious matter.

But before computing the level of false accepts, ask about the level of false accepts that is acceptable. The preferred answer is zero, but this is not achievable. Unfortunately, establishing a value for the acceptable level of false accepts involves many difficult issues. For a process that has been operating successfully for some time, one approach is to determine the level of false accepts with the procedures currently in use. Just knowing this number, however, can also have some side effects.

The level of false accepts depends on the following two ratios:

- The ratio of the process/product tolerance to the standard deviation for the process/product variations from target.
- The ratio of the process/product tolerance to the measurement uncertainty.

If the process/product tolerance is twice the process/product standard deviation and the process/product tolerance is four times the measurement uncertainty, the level of false accepts is approximately 0.8%. As either ratio increases, the level of false accepts decreases. However, there is a cost associated with increasing either ratio.

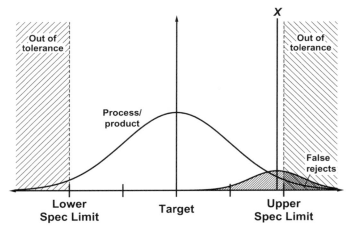

Figure 1.21. False rejects.

False Rejects Figure 1.21 depicts a situation in which the true value is slightly within the upper specification limit. The process is actually in tolerance. But because of the measurement uncertainty, the measurement result could be on the unacceptable side of the upper specification limit. For products, this leads to what is referred to as false rejects. An increase in the level of false rejects has the same consequences on process economics as a reduction in yield.

The level of false rejects also depends on the following two ratios:

- The ratio of the process/product tolerance to the standard deviation for the process/product variations from target.
- The ratio of the process/product tolerance to the measurement uncertainty.

If the process/product tolerance is twice the process/product standard deviation and the process/product tolerance is four times the measurement uncertainty, the level of false rejects is approximately 1.5%. As either ratio increases, the level of false rejects decreases. Again, there is a cost associated with increasing either ratio, but the economic benefits from reducing the level of false rejects just might offset these costs.

Accuracy Ratio

In the United States, a detailed analysis of measurement decision risk is usually avoided by relying on the accuracy ratio:

$$\text{Accuracy ratio} = \frac{\text{Process/product tolerance}}{\text{Measurement device accuracy or uncertainty}}$$

The objective of this approach is to ensure that measurement device errors are insignificant and may be ignored. The usual criterion for this is an accuracy ratio of 4:1, as in the case presented previously. For most applications, this accuracy ratio gives acceptable levels of false accepts (less than 1%) and false rejects (about 1.5%).

Figure 1.22 presents the distribution functions for process/product and measurement device for three cases:

Accuracy ratio of 8:1. The measurement device performance is better than necessary, which raises the potential for cost reductions.

Accuracy ratio of 4:1. This is the commonly desired accuracy ratio.

Accuracy ratio of 2:1. A detailed analysis of the measurement decision risk is recommended.

Guardbands

Figure 1.23 illustrates guardbanding, which is also known as *error budgeting*. The process/product specification limits are the same as before. But with guardbanding, the accept or reject decision is based on test limits instead of the specification limits. The tolerance for the test limits is smaller than the tolerance for the specification limits. The range of values between the test limit and the specification limit is the guardband. Guardbanding reduces the level of false accepts, but at the expense of increasing the level of false rejects.

Suppose a process temperature is to be $150°C \pm 10°C$. The temperature measurement device has a measurement uncertainty of $3°C$, giving an accuracy ratio of 10:3. If the level of false accepts must be 1% or less, we have three options:

- Increase the specification limits to $150°C \pm 12°C$. Such changes are more likely to be possible for process specifications than for product specifications, which are often set by the marketplace. But when the specification is on conditions within the process, this option should be put on the table. Such specifications are frequently set tighter than are really necessary.
- Install a temperature measurement device with a measurement uncertainty of $2.5°C$ or less.
- Apply guardbanding. Using a guardband of $3°C$, we accept temperatures of only $150°C \pm 7°C$. Although the guardband is typically close to the measurement uncertainty, the guardband for a 1% level of false accepts can be calculated.

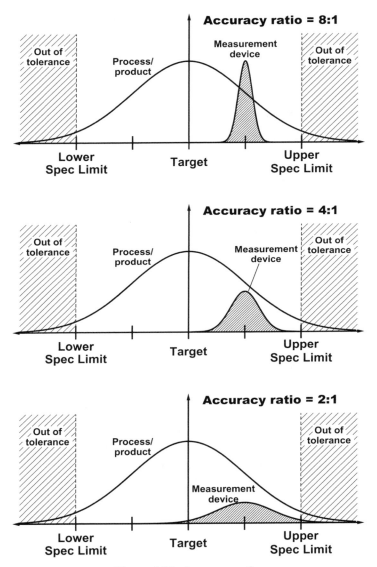

Figure 1.22. Accuracy ratio.

False Accepts with Guardbands Figure 1.24 depicts the use of a guard-band when making an accept or reject decision based on a measurement with a significant uncertainty. For a true value slightly outside the acceptable range, the false accepts are also depicted. The level of false accepts is lower with the guardband (Fig. 1.24) than it is without the guardband (Fig. 1.20).

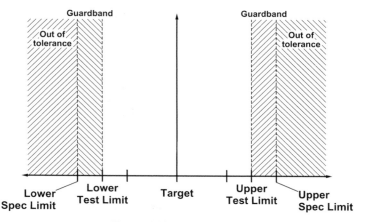

Figure 1.23. Guardbands.

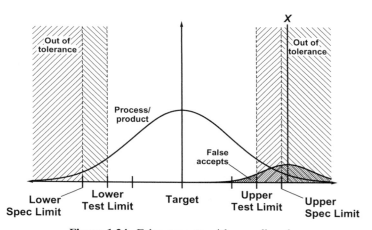

Figure 1.24. False accepts with guardbands.

Although the guardband is frequently set equal to the measurement uncertainty or accuracy, a more fundamental approach is as follows:

1. Establish an acceptable value for the level of false accepts.
2. Perform an analysis of the measurement decision risk to determine the value for the test limit that gives this level of false accepts.

As the measurement uncertainty increases, the magnitude of the guardband increases, and the test limit approaches the target. If the guardband is set equal to the measurement uncertainty, then for an accuracy ratio of 1:1 (measure-

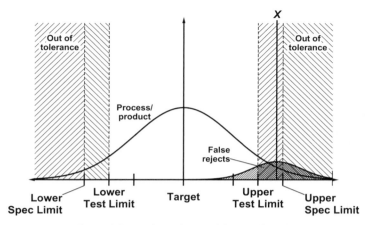

Figure 1.25. False rejects with guardbands.

ment uncertainty is equal to the process/product tolerance), the test limits are equal to the target. This situation is unacceptable.

False Rejects with Guardbands   Nothing comes without a cost. The consequences of guardbands are the following:

- The level of false accepts decreases. This is the positive aspect of guardbands.
- The level of false rejects increases. This is the negative aspect of guardbands.

Increasing the level of false rejects has the same impact on production economics as a decrease in yield. This is the source of much of the controversy surrounding the use of guardbands.

Figure 1.25 depicts the use of a guardband when making an accept or reject decision based on a measurement with a significant uncertainty. For a true value that is barely within the acceptable range, the false rejects are also depicted. The level of false rejects is larger with the guardband (Fig. 1.25) than it is without the guardband (Fig. 1.21).

1.11. CALIBRATION

The responsibility of maintaining standards for various quantities (mass, length, time, etc.) rests with governmental agencies. In the United States, the agency is the NIST. Counterparts exist in Great Britain, France, Germany, and Russia.

Calibration must always assess the performance of the measurement device by comparing values from the measurement device to values indicated by the standards. The two common ways of doing this are

- Applying the measurement device to samples for which a standard value has been previously established.
- Comparing the values from the measurement device to values from another device that is known to be in close agreement with the standard values.

In some but not all cases, calibration encompasses making adjustments to the measurement device to improve the agreement between the measurement device and the standard values. Some measurement devices, such as glass stem thermometers, provide no such adjustments. In industrial measurement devices, a common practice has been to adjust the zero and/or span to provide the best agreement between the values from the measurement device and the standard values. But today, the more common alternative is simply to replace an inexpensive measurement device or to return an expensive measurement device to the supplier for recalibration and/or repair.

Calibration Lab

In years past, most calibrations were performed by instrument technicians or other local personnel. For calibrating pressure measurement devices, almost all instrument shops routinely used a dead weight pressure tester (a weight placed on a hydraulic cylinder produces a known pressure). The zero and span of the pressure transmitter were adjusted so that the values indicated by the transmitter were in agreement with the values from the dead weight test equipment.

Changes in technology are making such practices less common. As delivered, the microprocessor-based pressure transmitters provide sufficient accuracy for most process applications. When required, the manufacturer will provide a calibration certificate. The digital transmitters are also far more stable, especially in regard to zero drift. If an inexpensive device such as a pressure transmitter is suspect, replacing the device is both easier and less expensive than recalibrating it.

For those devices that do require calibration, it is usually easier to retain the services of an outside organization. Calibration laboratories provide such services for a range of measurement devices, either on a measurement device shipped to their site or by bringing their equipment to the plant site. Many suppliers of measurement devices also provide calibration services for their products. Today, calibration is not just doing the technical work; it also entails generating the necessary documentation for review by others.

Traceability

To obtain ISO 9000 certification, the measurement devices that affect product quality must be identified. These devices must be calibrated at appropriate intervals using equipment whose relationship to the national standards is known.

NIST provides calibration services, mainly to calibration labs and manufacturers of test equipment. NIST will calibrate a measurement device or will determine the measured value for an artifact, reporting both relative to their standards with appropriate documentation. This enables a test equipment supplier to provide a document that states the relationship between the measured values from one of its products to the device that it submitted to NIST for calibration.

The result is a chain of calibrations that starts with the national standard and ends with the calibration of the industrial measurement device. A typical calibration chain is as follows:

National Institute of Standards and Technology

⇑

Test meter manufacturer's calibration facility

⇑

Test meter at commercial calibration laboratory

⇑

Industrial measurement device

Traceability simply means that every link in the chain is known and documented. There can be any number of links in the chain, but obviously there is some incentive to minimize the number of links.

Uncertainty

Each link in the traceability chain adds some uncertainty to the assessment of the performance of a measurement device. The test accuracy ratio (TAR) is the ratio of the accuracy of the meter or artifact on which the calibration is based to the accuracy of the measurement device being calibrated. Traditional practice was to require a ratio of 4:1 or better.

The traditional approach is being replaced by approaches based on quantifying the uncertainty associated with each calibration in the traceability chain. IEC 17025, *General Requirements for the Competence of Testing and Calibration Laboratories,*[5] is currently the ultimate authority for calibration standards and procedures. IEC 17025 also specifies the documentation (including uncertainty data) that must accompany a calibration. Most calibration labs are now IEC 17025 accredited. Accreditation is making possible numerous mutual recognition agreements (MRAs), whereby a calibration performed by a lab in one country is recognized by other participating countries.

Check Standard

Some measurement applications are amenable to check standards; others are not. The following are two examples of check standards:

- Gas sample of known composition to be analyzed periodically by an analytical instrument.
- Sheet of known weight or thickness to be sensed by a weight or thickness gauge.

These examples are considered to be artifacts for which the expected results of the measurement are known with much less uncertainty that that associated with the measurement device. The use of a check standard permits data to be collected that can expose any deterioration of the measurement device with time.

Check standards are relatively easy to incorporate into the procedures for sampling analyzers such as gas chromatographs. To provide the check standard, a cylinder is prepared with a gas of known composition that is typical of the process samples (there are commercial organizations that specialize in supplying such cylinders). On a specified time interval, a sample from the cylinder is injected into the chromatograph in lieu of the process sample. Whether this is once an hour, once a day, or once a week has to be established based on prior experience on the specific analysis to be performed.

In sheet processing applications such as paper machines, the weight or thickness of the sheet is measured by a gauge that scans across the sheet. Between scans, the gauge can be positioned over a check standard so that the gauge can sense its weight or thickness.

1.12. MEASUREMENT DEVICE COMPONENTS

Figure 1.26 presents a schematic of the physical components of a measurement device. We want to examine the role of each.

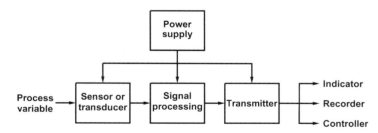

Figure 1.26. Components of a measurement device.

Sensor or Transducer

The process variable influences some characteristic of this component that can be used as the input to the signal processing components. Most rely on some basic physical principle. Examples are the following:

- Temperature affects the resistance of a noble metal such as platinum (RTD).
- The flow through an orifice affects the pressure drop across the orifice (orifice flow meter).
- The force on a displacer is the weight of the displacer less the weight of liquid being displaced (displacer level gauge).
- A conductor moving through a magnetic field generates a potential that is proportional to the velocity of the conductor and the strength of the magnetic field (magnetic flow meter).

Signal Processing

The objective of the signal processing components is to generate a signal that is linearly related to the process variable. The characteristic of the sensor that is affected by the process variable is usually not linearly related to the process variable. But because the nature of this relationship is known, the signal processing components can be designed so that the output from the signal processing components is linearly related to the process variable—that is, one function of the signal processing components is linearization.

For some measurement devices, other functions are required. One is to compensate for other variables that affect the measured value. For example, when using a thermocouple to measure process temperature, the output must be compensated for changes in the temperature of the reference junction.

Analog Transmitter

In electronic measurement devices, the output of the signal processing components is often a 0- to 1-V or a 0- to 5-V signal. If such a signal extends over any significant distance within an industrial facility, problems such as the following arise:

- The magnetic fields associated with power equipment affect the voltage levels within such signals.
- The voltage drop resulting from current flowing through the signal wires introduces errors into the signals.

Consequently, a medium that is not affected by such factors is required. For electronic signals, the medium normally chosen is current. An electronic transmitter converts the voltage signal into a current signal that is not affected by magnetic fields and resistances in the wires that carry the signal.

Digital Transmitter

The advantages of microprocessor technology for signal processing led to the development of the so-called smart transmitter. The capabilities of the microprocessor enable the signal from the basic sensor to be more accurately converted to a value in engineering units. More accurate linearization relationships are employed, and the signal from the sensor is compensated for the influence of temperature, pressure, etc.

To date, most smart transmitter products have retained the capability to transmit the measured value in analog form via a current loop. However, the use of current loops for signal transmission is gradually being replaced by digital transmission via a communications network. More on this shortly.

Power Supply

The signal processing and transmitter components of the measurement device always require a source of power. Some sensors require power, but many do not. For example, thermocouples generate a millivolt electrical potential that is a function of the difference in temperature between the two junctions. No external power is required.

The power supply is often not physically within an electronic measurement device itself. The measurement device must be located in the production area. In some facilities, this raises the issue of the presence of explosive vapors, which requires either of the following:

- Proper enclosures for the measurement device.
- Intrinsically safe designs of the electronic circuitry.

When the latter is pursued, advantages accrue from locating the power supply in a remote area where it will not be exposed to the explosive vapors.

1.13. CURRENT LOOP

At the turn of the millennium, the most common method for transmitting the measured value from the measurement device to a controller, computer, recorder, and so on, was via the 4- to 20-ma current loop illustrated in Figure 1.27. The correspondence between the current loop values and the measurement range is as follows:

$$4\,ma(0\%) \Leftrightarrow \text{Lower-range value(zero adjustment)}$$

$$20\,ma(100\%) \Leftrightarrow \text{Upper-range value(zero adjustment + span adjustment)}$$

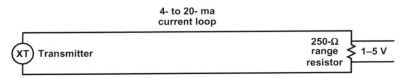

Figure 1.27. Current loop for signal transmission.

The current loop is said to provide a *live zero*—that is, an input of 0% of span is a current loop value of 4 ma. An input of 0 ma does not correspond to a valid measured value; in fact, this input means a failure of some type. It could possibly be due to loss of power or some other failure at the transmitter. It could also result from a disruption, or lack of continuity, in the circuit itself. Most input systems provide *open circuit detection* logic, which means that an input of zero is reported as an error.

Why Use Current?

In most industrial facilities, the measurement device is physically located at some distance from the controller, computer, recorder, etc. It may be less than 100 ft in small facilities, but could be over 1000 ft in large facilities such as refineries and paper mills. Two issues arise for voltage transmission:

- Some current will flow in the circuit, which results in a voltage loss due to the resistance of the wires.
- Most facilities contain power equipment, mainly electric motors, that generate magnetic fields of varying intensity. Such magnetic fields can lead to an electric potential—usually called an *induced voltage*—in electrical circuits. Good wiring practices, specifically shielding, can reduce the induced voltages to small values.

A major advantage of using a current loop for signal transmission is that the current flow is immune to these factors. Anything that influences the current flow around the circuit will affect the current flow at both the transmitter and the receiver. The transmitter must continuously monitor its current output and make the necessary adjustments to maintain the desired current flow. The performance of the current loop for signal transmission depends solely on how well the transmitter does this. The current flow at the receiver is exactly the same as the current flow at the transmitter.

Range Resistor

Current is used for signal transmission only. The internal workings of most electronic devices are based on voltages. Electronic analog (conventional) transmitters first represent the measured value as a voltage signal, which is subsequently converted to a current signal for transmission.

At the receiver end, the current signal can be converted to a voltage signal using a resistor, usually referred to as the *range resistor*. Figure 1.27 illustrates the simplest approach, which involves inserting a 250-Ω range resistor into the circuit to obtain the following conversion:

$$4\,\text{ma}(0\%) \Rightarrow 1\,\text{V}$$

$$20\,\text{ma}(100\%) \Rightarrow 5\,\text{V}$$

It is possible to have more than one range resistor in the circuit. This was common in conventional electronic systems, with one range resistor for the controller and a separate one for a display. Even some early digital systems inserted one range resistor for the input to the computer and a separate range resistor for the analog backup system. The maximum number of range resistors is determined by the maximum impedance in the current loop permitted by the transmitter. Today, most current loops contain only one range resistor—for the input to a digital system of some type.

More complex circuits can convert the current loop signal into a 0- to 5-V signal. For example, applying a −1.25-V bias to a 312.5-Ω range resistor converts 4 ma to 0 V and 20 ma to 5 V. This has the advantage of using the entire input range of the A/D converter. However, we will consider only the simple approach.

A/D Converter

As illustrated in Figure 1.28, the voltage across the range resistor is the input to an A/D converter, which converts the voltage input to a digital value. The resulting digital value is a short integer (16-bit) value that is generally referred to as the *raw value*. The input processing hardware and associated software direct the A/D converter to convert the input voltage to a digital value and store the result in the proper location in the *raw value table*.

The two key parameters for the A/D converter are

- Input voltage range.
- Resolution.

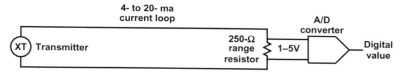

Figure 1.28. Current loop with A/D converter.

Other specifications apply, such as the time interval between conversions. But because most process applications are very slow, these are rarely an issue.

Input Voltage Range

For process applications, most A/D converters are 5-V converters. However, there are a couple of details that require explanation. Most current loop transmitters can output values somewhat over 20 ma. The exact overrange varies from one manufacturer to another. Most are on the order of 5% of the 16-ma span, giving a maximum value of 20.8 ma, or 5.2 V. Some 5-V A/D converters will process input values up to 5.12 V, which is a convenient value ($512 = 2^9$) for binary numbering systems. Although this may not accommodate the entire overrange available from the transmitter, it is sufficient.

A/D converters may be either unipolar (positive voltages only, or 0 to +5 V) or bipolar (positive and negative voltages, or −5 to +5 V). For current loop inputs, a unipolar A/D converter is sufficient. However, some systems will accept direct thermocouple inputs (after amplification). Because signals from thermocouples may be positive or negative, a bipolar A/D converter is required.

Resolution

Because most A/D converters generate binary numbers, the resolution is specified as the number of significant bits in the binary value. However, there is one subtle aspect that must be understood. For example, a 12-bit A/D converter provides a resolution of 1 part in $2^{12} = 4096$. This is applied to the input voltage range, which depends on whether the converter is unipolar or bipolar:

Unipolar, 12-bit converter. Input range is 0 to +5.12 V; resolution is 5.12/4096 = 1.25 mv.

Bipolar, 12-bit converter. Input range is −5.12 to +5.12 V; resolution is 10.24/4096 = 2.5 mv

In effect, a 12-bit bipolar A/D converter provides 11 data bits plus a sign bit, which is sometimes explicitly stated as *11 data bits plus sign.*

For process applications, most manufacturers choose A/D converters with 11 ($2^{11} = 2048$) or 12 ($2^{12} = 4096$) data bits. It is possible to obtain converters with 14 ($2^{14} = 16384$) or 15 ($2^{15} = 32768$) data bits. But unless the source of the input signal is extremely stable, the additional bits will be mostly noise.

Effective Resolution

If the input range is 0 to 5.12 V, an 11-bit unipolar A/D converter provides a resolution of 2.5 mv, which is 0.0488% of the input range. But when used for current loop inputs as in Figure 1.28, the resolution should really be expressed as a percent of the usable input range, which is 1 to 5 V for the configuration in the figure.

For current loop inputs configured as in Figure 1.28, using an 11-bit A/D converter with an input range of 0 to 5.12 V provides the following performance:

A/D converter input range	0 to 5.12 V
A/D converter resolution	11 bits (2.5 mv or 1 part in 2048)
Resolution for input range	2.5 mv or 0.05% of a 0- to 5-V input range
A/D converter raw value range	0 to 2000 for a 0- to 5-volt input range
Effective input range	1 to 5 V
Effective resolution	2.5 mv or 0.0625% of a 1- to 5-V input range
Effective raw value range	400 to 2000 for a 1- to 5-V input range
Effective resolution in bits	1 part in 1600 or 10.6 bits ($\log_2 1600 = 10.6$)

For a temperature transmitter with a measurement range of 50° to 250 °F, the resolution in temperature units is 0.0625% of 200 °F or 0.125 °F. The repeatability of measurement devices such as coriolis flowmeters and RTDs can be 0.0625% or better, and thus the argument for A/D converters with 12 data bits. Especially in applications such as laboratories and pilot plants where the quality of the data is paramount, the resolution of the A/D converter deserves some attention.

Amplifiers and Multiplexers

Most input systems contain a few components in addition to those illustrated in Figure 1.28. To provide a stable input to the A/D converter, an amplifier is usually inserted between the voltage inputs from the range resistor and the A/D converter. For current loops, the amplifier does not alter the voltage level. But when used in applications such as direct thermocouple inputs, the amplifier increases the voltage level from the thermocouple's millivolt level to the 5-V range required by the A/D converter.

Most analog input cards provide multiple analog inputs, the number typically being in the range of 4 to 16 (computer people do like those powers of 2!). Some cards provide an A/D converter for each input; others use a multiplexer to switch the inputs onto a single A/D converter. This decision affects the input scan software provided by the manufacturer, but otherwise there should be no effect on users of the system.

1.14. POWER SUPPLY AND WIRING

There is one very important component missing from the current loop shown in Figure 1.28: a source of power. To date, this seems to require two strips of copper—that is, wires—between the source of power and the transmitter. Perhaps battery-powered transmitters will someday become the norm, but so far there has been little use of such transmitters.

Depending on the power requirements of the transmitter, the wiring configuration will be one of the following:

- Two wire.
- Three wire.
- Four wire.

Two-Wire Transmitters

As illustrated in Figure 1.29, in two-wire transmitters the data transmission and the power share the same two physical wires. The source of power may be integrated into the input-processing card, or may be provided by a separate power supply (as shown).

The power supply is typically a 24-V DC source of power. Suppose the current flow is at the lower range value of 4 ma. If the full 24-V were taken across the transmitter, the available power at the transmitter would be

$$\text{Power} = (4\,\text{ma})(24\,\text{V}) = 0.096\,\text{W}$$

The practical limit is somewhat less, but this is adequate for low-power measurement devices such as those for temperature and pressure.

When digital transmission is used, the digital transmission can also share the same two physical wires that are used to power the transmitter. The requirement for power reduces the attractiveness of technologies such as fiber optics and wireless. As long as you have to install physical wires to the transmitter for power, the possibility exists to use this same pair of wires for data transmission as well.

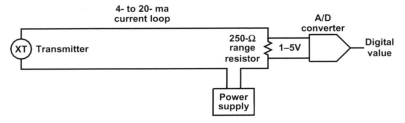

Figure 1.29. Two-wire transmitter.

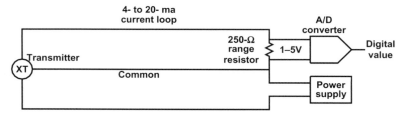

Figure 1.30. Three-wire transmitter.

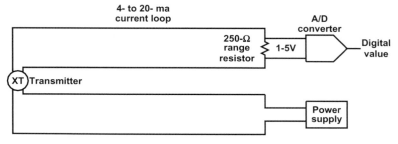

Figure 1.31. Four-wire transmitter.

Three-Wire Transmitters

In the three-wire transmitter shown in Figure 1.30, separate circuits are provided for the current loop and the power supply, but they share a common return, which is normally at ground potential. The power supply is always a DC supply that is separate from the input card. This configuration is typically used for transmitters that require a moderate amount of power.

Four-Wire Transmitters

Figure 1.31, illustrates a four-wire transmitter with separate circuits for the current loop and the power supply. There are no common elements between the two circuits. The power supply is usually AC, but could be DC (if DC, the three-wire configuration can usually be used). The power supply is always separate from the input card. There is no limit on the amount of power that can be supplied to the transmitter using this configuration. It is the choice for all transmitters with high power requirements and for all transmitters that require AC power.

1.15. SERIAL COMMUNICATIONS

Figure 1.32 illustrates using serial communications to transfer information from the transmitter to a digital system (such as a control system). The trans-

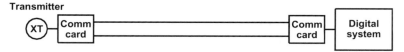

Figure 1.32. Signal transmission via serial communications.

mitter and the digital system are equipped with serial communications cards that provide one or more "ports" for serial interfaces. When the connection, or *link*, is from a port on the transmitter to a port on the digital system, the communications structure is said to be *point to point*.

The communications cards on a given link must be compatible. The common types are

RS-232. Most common, but limited to short distances (50 ft) and provides no electrical isolation between transmitter and receiver—that is, both must have same ground, which is a problem for industrial installations.

RS-485. Capable of longer distances and provides electrical isolation so that a common ground is not required. Consequently, RS-485 is more suitable for industrial installations.

If compatible cards are not available, modules that convert between RS-232 and RS-485 can be purchased.

Protocol

RS-232 and RS-485 are hardware standards only. That is, they specify the number of wires, the use of each wire, the voltage levels, etc. This permits binary information to be transmitted from the transmitter to the digital system but says nothing about the content. The protocol specifies the content of the transmission. For example, if the transmitter is to transmit a value to the digital system, the protocol specifies the exact sequence of characters to be transmitted.

A transmission could be entirely text information, such as the character string *127.8* followed by a carriage return and line feed (the *terminator*). Some transmitters can be configured for *continuous transmission*, by which they transmit the current value to the digital system on a fixed time interval. However, this is a very simple protocol, is one-way only (transmitter to digital system), can be used only in point-to-point configurations, and does not use the full potential of the serial communications interface. Therefore, more elaborate protocols are often used.

Software

A software module is required to convert the string of characters or bytes received from the transmitter to a numerical value for the measured variable.

Developing such software modules is complicated by the fact that there is no effective standard for the protocol. This has led some manufacturers to develop their own specifications for the protocol, resulting in a so-called proprietary protocol.

With some experience, such software modules can be developed quite quickly. However, experience has shown that these modules lead to support problems. Suppose one upgrades either the transmitter side of the link or digital system side of the link. Will the software module for the protocol still work? Usually it does, but occasionally it does not. Manufacturers have also been known to change the protocol during the life of a product. This would mean that the transmitters now being purchased use a different protocol from those previously installed. Basically, the software module for the protocol is a constant source of potential problems.

In industrial applications, the software module is commonly called a *driver*. However, this use of the term is not entirely consistent with its use in the context of computer operating systems.

Modbus

Developed in the 1970s by Modicon (a manufacturer of PLCs) for transferring information between PLCs or between a PLC and a host computer, Modbus is about the nearest thing to a standard for a protocol.[6] But even it is subject to extensions that, if used, will require customization of the software module. The original Modbus protocol is summarized as follows:

Byte	Contents
1	Start of transmission (STX) (02 hex)
2	Function code
3	Number bytes of user data (N)
4 to $N + 3$	User data (depends on function code)
$N + 4$	End of transmission (ETX) (03 hex)
$N + 5$ to $N + 6$	Cyclical redundancy check (CRC)

The second byte in the Modbus protocol is a function code. Modicon defined function codes for PLC applications, such as reading registers and writing registers. Some transmitters use these function codes. However, Modicon left a number of function codes undefined so that other users of the protocol could extend it. Some manufacturers of measurement devices have chosen to do so, often for good reason (measurement devices are quite different from PLCs). If the manufacturer has extended the Modbus protocol by defining function codes specific to its product, a standard Modbus software module would not support such function codes.

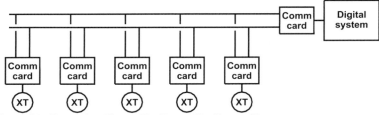

Figure 1.33. Multidropping.

Multidropping

Using point-to-point serial links would require the digital system to provide a serial port for each transmitter. This could lead to a large number of serial ports. For some computers, large numbers of serial ports are readily available, but others have severe limitations. The alternative is to use the multidropped arrangement illustrated in Figure 1.33.

To do so requires RS-485 communication cards for the hardware and a software module that can *poll* the individual transmitters. Each transmitter is assigned an address. The communications protocol must provide a field for the address of the transmitter (note that the standard Modbus protocol does not, although one could be easily added). To obtain the current value of the measured variable from a transmitter, the software in the digital system must issue a *read current value* request to the appropriate transmitter, which responds with the current value of its measured variable. This results in a slower scan rate for the measured variable, but 1-sec scans can normally be achieved with multidropped configurations containing eight or fewer transmitters.

Mass Flow Meters

Serial communications is not the interface of choice for most transmitters. However, suppose a measurement device produces more than one measured value. A good example of such a measurement device is a mass flow meter. The following measured values are available:

- Mass flow.
- Temperature.
- Density.
- Viscosity (if this option is purchased).

If a serial interface is used, all four values could be retrieved over the serial link. Using current loop technology, four current loops would be required. To make matters more complicated, one commercial supplier of mass flow meters provided only two current loop outputs in its products. The unit could be configured to transmit any of the listed values over a given current loop. However, the net result is that only two of the values could be retrieved using current loops.

Composition Analyzers

Composition analyzers such as chromatographs are sampling measurement devices. An analysis is generated at discrete intervals of time, depending on the characteristics of the separation column within the chromatograph. Furthermore, each analysis usually consists of multiple values, one for each component. In addition, the time that the sample was injected is also of interest.

Current loop interfaces do not handle such analyses very well. Occasionally the analysis can be reduced to the ratio of two key components, which can be transmitted using a current loop. But even in such situations, the time of the last analysis cannot be captured this way.

Weight Transmitters

Load cell technology is capable of resolutions far beyond what is practical with current loops. A 15-bit (data) A/D converter provides a resolution of 1 part in 32,768. Load cells are available with five-digit local displays, which provides a potential resolution of 1 part in 100,000. For example, the load cell that can display a weight from 0.0 to 6,000.0 kg with a five-digit display has a display resolution of 1 part in 60,000. Such a resolution cannot be achieved with current loop technology.

Suppose a 12-bit A/D converter with an input range of 0 to 5.12 V is used to process a current loop input from a load cell that displays weights up to 6,000 kg to a resolution of 0.1 kg (a five-digit display). Assuming a 250-Ω range resistor is inserted into the current loop circuit, a weight of 0.0 kg gives a raw value of 800. A weight of 6,000.0 gives a raw value of 4,000. A change of 1 count in the raw value corresponds to a change in weight of 6,000.0/3,200 = 1.9 kg. Whereas the load cell is capable of indicating a weight change of 0.1 kg, the smallest weight change that can be detected from the input signal is a change of 1.9 kg. Consequently, the value displayed by the digital system rarely agrees with the value in the local display of the weight transmitter.

1.16. SMART TRANSMITTERS

The marketing types have embraced the term *smart* to the extent that everything is now advertised as "smart"—even consumer products such as ovens and washing machines. The common denominator seems to be that a microprocessor has been somehow incorporated into the product. The predecessors to the smart transmitters were electronic analog transmitters. These are now usually referred to as *dumb* transmitters; obviously, nobody wants these anymore.

The microprocessor does indeed provide the opportunity for designers to incorporate some very useful features into the measurement device. The following list summarizes some of the possibilities:

- Checks on the internal electronics.
- Checks on environmental conditions, such as temperature, within the measurement device.
- Compensation of the measured value for conditions within the measurement device.
- Compensation of the measured value for other process conditions.
- Linearizing the output of the transmitter.
- Configuring the transmitter from a remote location.
- Automatic recalibration of the transmitter.

These features do indeed give the smart transmitters clear advantages over their predecessors.

Although not an explicit feature, a major advantage of smart transmitters is their versatility. Using the same model of a measurement device in multiple locations reduces spare parts requirements. If the requirements of the application change, a more versatile measurement device is likely to accommodate such changes. This is a major advantage in laboratory and pilot plant applications.

Remote Configuration

A smart transmitter can be configured remotely—that is, it is not necessary to go to the physical location of the transmitter, open the enclosure, and make adjustments. There are three options:

- A handheld battery-powered microprocessor-based configuration unit. Instrument technicians seem to like these and become very adept with them, but you may be surprised at their cost.
- Centralized device management software that maintains a database of the configurations. Usually this is integrated with the configuration software for the control system.
- PC-based device configuration software. Sometimes, but not always, the device management software part of the control system configuration software can be extracted and used in a stand-alone manner.

Remote configuration definitely facilitates maintenance—replace the measurement device and then load the configuration from either the handheld unit or the central database. This can be done rapidly and free of errors (provided the correct configuration data set is downloaded into the replacement).

One of the issues with remote configuration tools is that they should support products from multiple vendors. This is accomplished via device descriptions or profile descriptions, the most common being the following:

- Electronic device descriptions (EDDs).
- Profibus-process automation (profibus-PA) profile description.
- Highway Addressable Remote Transmitter (HART) device description.

The supplier of the device management software provides the descriptions available at the time the software was delivered. For later products, the manufacturer of the measurement device provides the description in the form of a software module that can be downloaded via the Internet.

HART can be used to communicate with devices connected via a 4- to 20-ma current loop. A high-frequency signal (1200 Hz represents binary 1; 2200 Hz represents binary 0) with an amplitude of 0.5 V or less is superimposed on the same two wires used for the current loop. A high-pass filter separates the communications signal; a low-pass filter separates the 4- to 20-ma signal.

Accuracy

One area that smart transmitters deliver better performance then do older models is in regard to accuracy. There are two aspects of this:

- The initial accuracy specifications are normally superior.
- The drift (change of parameters with time) is less.

Although complete automatic recalibration is rarely provided, many smart transmitters can automatically reset their zero adjustment. In conventional transmitters, zero drift contributed to a deterioration of the accuracy with time. When the zero can be reset automatically, this source of error is greatly reduced if not completely eliminated.

Interfaces

Being digital in nature, smart transmitters provide three options for interfacing with computers, control systems, etc.:

- Current loop.
- Network interface (usually referred to as *fieldbus*).
- Serial communications interface.

Almost all smart transmitters have current loop and fieldbus interfaces; the serial communications interface is less common. The current loop interface is

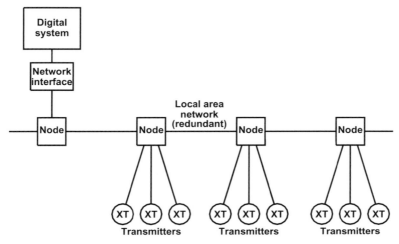

Figure 1.34. Fieldbus.

expected to gradually disappear as fieldbus networks become more widely installed.

Fieldbus

Fieldbus is basically a local area network (LAN) designed for communicating between field devices, including but not limited to smart transmitters, and digital systems such as controllers. Fieldbus uses LAN technology for communications and thus is flexible, two-way, capable of reading multiple values from a single transmitter, and so on.

Figure 1.34 presents a simplified structure for fieldbus. LAN technology is used to transfer information between the various nodes on the fieldbus network. The master node interfaces to the digital system, which provides data acquisition, process control, or other functions. The other nodes are slave nodes that communicate with the individual field devices, such as the transmitters.

Although most manufacturers seem to focus on reduced wiring costs as the major incentive for fieldbus, the interface's capabilities and flexibility prove far more advantageous. As compared to current loop installations, the initial commissioning of the system is much easier and can be accomplished in less time. Thereafter, addition of a new measurement device is greatly simplified.

In the process industries, fieldbus got off to a slow start. As is customary with digital technology, several designs for fieldbus appeared, all of which could meet the needs of process applications. The delays largely revolved around standards (or lack thereof) for the communications network:

- Adopting a standard for the protocol became embroiled in international politics. Although a standard is now available,[7] it is in reality *five* standards rolled into one document. That is, the standard provides five alternatives for fieldbus implementations. The marketplace will ultimately determine the winner.

- Integrating digital transmission into control systems required substantial systems efforts on the part of control system suppliers. Many suppliers were naturally reluctant to make such an investment until the standards issue was resolved.
- Without a standard, users were hesitant to install equipment that might be incompatible with the eventual standard.

The following process installations are encountered most frequently:

- Profibus, which appears to be prevailing in Europe.
- Foundation fieldbus, which appears to be prevailing in the United States.

Most manufacturers of measurement devices can supply a model for either of these (in addition to a model for current loop installations and possibly a model for serial communications).

1.17. ENVIRONMENTAL ISSUES

Environmental issues pertain to the electronics associated with the measurement device. These are not directly exposed to the process materials but instead are exposed to the ambient conditions that surround the location of the measurement device. The following must always be considered:

Temperature. This is the temperature at the location in which the electronics are physically located.

Atmosphere. The atmosphere surrounding the electronics can raise a number of issues.

Procedures. Let's illustrate with an example. In food-processing facilities, a crew regularly cleans up the area, which usually involves high-pressure water hoses or the like that could be directed at the measurement device.

Ambient Temperature

Ambient temperature applies to the signal processing and transmitter components of the measurement device. The manufacturer of a measurement device will state actual values, but typical limits are $-25°$ to $+80°C$. Sometimes the term *noncondensing* is appended; electronic equipment performs better when dry.

The physical location of the sensor or transducer is determined by the process variable that is the subject of the measurement. But for the components that provide the signal processing and the transmitter, the options are

Integrally mounted. The signal processing and the transmitter are within the same physical enclosure as the sensor or transducer.

Remotely mounted. The signal processing and the transmitter are in an enclosure that is separated by some distance from the physical location of the sensor or transducer. In most situations, this entails some distance of low-level (millivolt) signal wiring, which raises shielding, grounding, and other such issues.

The choice is often dictated by the process temperature at the location in which the measurement is required. Where the process temperature is elevated, active cooling can be considered. One approach is to provide a constant purge of instrument air through the enclosure containing the transmitter. A more extreme approach is to provide a water jacket around the enclosure. The downside of active cooling is that the air purge or water flow through the jacket must be maintained at all times. This always proves to be more difficult than one initially suspects; even a brief loss of air purge or water flow usually has adverse consequences.

Enclosures

Enclosures provide protection. The electronics within a measurement device are almost always protected from various elements of the environment by an appropriate enclosure. The requirements depend on the application, but the common requirements include the following:

Solids. Protect electronic equipment from foreign objects (or protect foreign objects such as hands and fingers from electronic equipment).

Liquids. Keep electronic equipment dry.

Corrosion. Protect electronic equipment from corrosive liquids or gases.

Explosive atmospheres. Either keep explosive atmospheres away from sources of ignition (purged enclosures) or contain the explosion in case of ignition (explosion-proof enclosures).

The manufacturer of a measurement device typically provides the enclosure, but it is the user's responsibility to make sure that the enclosure is appropriate to the environment within his or her facility.

The enclosure is accompanied by documentation that certifies its suitability for certain environments. These certifications are based on two standards:

NEMA Standard 250, a U.S. standard.

IEC 60529, a European standard.

Most other countries either use one of these directly or have a standard derived from one of these. Most manufacturers of enclosures can provide certifications appropriate for both standards.

NEMA Standard 250 The National Electrical Manufacturers Association (NEMA) is the U.S. manufacturer's organization that develops and maintains various standards. Also included are the test methods used by Underwriters Laboratories (UL) to verify that a product conforms to the standard.

NEMA Standard 250, *Enclosures for Electrical Equipment,*[8] defines 13 enclosure "types," some of which have letters appended to the type number to specify a variation of the basic type. Each type is briefly described as follows:

Type	Location	Brief Description
1	Indoor	General purpose
2	Indoor	Drip tight
3	Outdoor	Rain tight, sleet resistant
3R	Outdoor	Type 3 and dust tight
3S	Outdoor	Type 3R with external mechanism if ice laden
3X	Outdoor	Type 3 and corrosion resistant
3RX	Outdoor	Type 3R and corrosion resistant
3SX	Outdoor	Type 3S and corrosion resistant
4	Both	Watertight, dust tight
4X	Both	Type 4 and corrosion resistant
5	Indoor	Dust tight
6	Both	Submersible, watertight, and dust tight
6P	Both	Type 6 for prolonged submersions
7	Indoor	Explosion-proof (Class I, Groups A, B, C, D)
8	Both	Oil-immersed equipment (Class I, Groups A, B, C, D)
9	Indoor	Explosion-proof (Class I, Groups E, F, G)
10	Mines	Explosion-proof in methane or natural gas
11	Indoor	Corrosion resistant and dust tight, oil immersed
12	Indoor	Dust tight and drip tight
12K	Indoor	Type 12 with knockouts
13	Indoor	Oil tight and dust tight

This is intended only as a general guide. Consult the latest version of the standard for current and detailed information on NEMA types for enclosures.

For industrial measurement devices, enclosures generally have a NEMA 4X classification. Although this enclosure is corrosion resistant, the standard says nothing about what corrosive materials are resisted. NEMA 4X enclosures do not provide adequate protection in locations where explosive atmospheres may be present (see "Hazardous Area" in section 1.18).

IEC 60529   The International Electrotechnical Commission (IEC) develops and maintains international standards for electrical equipment and related technology. Through the European Commission on Electrotechnical Standardization (CENELEC), these standards are almost universally used in Europe.

IEC 60529, *Degrees of Protection Provided by Enclosures*,[9] defines ingress protection (IP) codes that are applied to enclosures. The format for the codes is IPxy, where x is a digit in the range 0 through 6 that pertains to protection from solid objects; y is a digit in the range 0 through 8 that pertains to protection against liquids. The following is a very brief description of the value of each digit:

Digit x	Protection from Solid Objects
0	No protection
1	Objects > 50 mm (hands)
2	Objects > 12 mm (fingers)
3	Objects > 2.5 mm (wires)
4	Objects > 1 mm (thin wires)
5	Dust (limited ingress)
6	Dust (totally protected)

Digit y	Protection from liquids
0	No protection
1	Vertically falling droplets
2	Sprays up to $15°$ from vertical
3	Sprays up to $60°$ from vertical
4	Sprays from all directions
5	Low-pressure jets of water
6	High-pressure jets of water
7	Low depth, brief immersion
8	Prolonged immersions

These lists are intended only as a general guide. Consult the latest version of the standard for current and detailed information on IP codes for enclosures.

For industrial measurement devices, enclosures generally have either an IP66 or an IP67 classification.

1.18. EXPLOSIVE ATMOSPHERES

For a flammable gas or vapor in air, the requirements for a fire or explosion are as follows:

- Concentration above the lower explosive limit (LEL).
- Concentration below the upper explosive limit (UEL).
- Source of ignition (spark, hot surface, etc).

The LEL and UEL depend on the specific component (or mixture of components) present in the air. The source of ignition also depends on the amount of energy needed to ignite the vapors.

Dust, fibers, and many other combustible solid materials in air can cause explosions that are more severe than flammable gases or vapors.

Hazardous Area

Any area in which flammable gases or vapors may be present continuously under normal operations, intermittently under normal operations, or only under abnormal operations is subject to a hazardous area classification. There are two options for hazardous area classifications:

- Class, division, and group as per NEC Article 500.
- Zone as per IEC 60079 or NEC Article 505.

Before any electrical equipment, including measurement devices, can be specified, the hazardous area classification must be established. The responsibility of the person specifying a measurement device for a given application is that the specifications be appropriate for the previously established classification of the hazardous area.

We will give only a brief introduction. The respective standards are the final word; but many of the suppliers also provide very informative and readable literature on the subject (for example, R. Stahl offers *Basics of Explosion Protection*[10] as a download from its website).

NEC Article 500   Developed by the National Fire Protection Agency (NFPA), the National Electric Code (NEC) is the governing standard on electrical safety within the United States. Article 500 pertains to the classification of hazardous locations,[11] and is summarized as follows:

Class. Pertains to the nature of the combustible material.
    Class I. Flammable gases or vapors.
    Class II. Combustible dust.
    Class III. Easily ignitable fibers.

Division. Pertains to the frequency at which the combustible material is present.

Division 1. Present during normal operations.

Division 2. Not normally present.

Group. Pertains to the combustible material.

Group A. Acetylene.

Group B. Hydrogen, ethylene oxide, etc.

Group C. Ethyl ether, ethylene, etc.

Group D. Gasoline, hexane, etc.

Group E. Metal dust.

Group F. Carbon black, coal dust, etc.

Group G. Flour, starch, etc.

Division 2 is less demanding than Division 1. When functioning properly, most measurement devices (there are exceptions) would not provide a source of ignition—that is, they do not spark, they have no hot spots, etc. Even if exposed to an explosive atmosphere, there would be no detonation. For Division 2, the detonation would occur only if two failures occurred simultaneously:

- Explosive vapors are present (infrequent for the Division 2 classification).
- The measurement device fails (also infrequent).

For Division 1, the assumption is basically made that explosive vapors will be present at the time the measurement device fails. Therefore, either an appropriate enclosure must be provided for the measurement device or the measurement device must be designed so that no failure can lead to a spark or other source of ignition (intrinsically safe designs, to be discussed later).

IEC 60079   Developed by the IEC, standard 60079 is the governing standard on classification of hazardous areas used throughout Europe.[12] This standard introduced the concept of zones:

Zone 0. Flammable gases or vapors very frequently present or continuously present, even under normal operations.

Zone 1. Flammable gases or vapors occasionally present under normal operations.

Zone 2. Flammable gases or vapors present infrequently or for short periods (such as during abnormal operations).

Zones 20, 21, 22 are the corresponding zones for dusts.

Beginning in 1996, similar classifications were incorporated into the NEC, eventually evolving to NEC Article 505. While very similar, these two classifications are not identical. The general relationship between zones and divisions is typically considered to be as follows:

- Division 1 of NEC Standard 500 encompasses Zones 0 and 1 of IEC 60079/NEC Article 505.
- Division 2 of NEC Standard 500 is equivalent to Zone 2 of IEC 60079/ NEC Article 505.

The splitting of Division 1 into two zones reflects the special issues that arise when explosive vapors are continuously present. When explosive vapors are present intermittently (Zone 1), the area can be monitored with a gas sniffer to make sure that no explosive conditions exist, which permits work to be performed on electrical equipment mounted within enclosures. But because explosive vapors are considered to be continuously present for Zone 0, this means that electrical equipment mounted within enclosures must be powered down to perform work. This usually means a plant shutdown.

Explosion-Proof Enclosures

Enclosures have been designed to contain any explosions. Flammable gases and vapors can enter such an enclosure, but should the mixture be explosive and be ignited by the electrical equipment inside, the enclosure is designed to contain the explosion in such a manner that it is not an ignition source for an external explosion. The surfaces of the enclosure must remain cool, and all escaping gases must be sufficiently cooled. Explosion-proof enclosures tend to be heavy.

The advantages and disadvantages of explosion-proof enclosures are summarized as follows:

- Sturdy, but expensive, enclosure.
- Well accepted in the United States, but not everywhere.
- Competent installers available, but installation cost high.
- When used in Zone 0, equipment within enclosure must be shut down for work to be performed.
- Only option for high-power electrical devices.
- Safe-area certification not required for electrical equipment.
- Cooling an issue for heat-generating electrical equipment.

From a measurement device perspective, cost often becomes an issue because the cost of the enclosure can easily exceed the cost of the measurement device itself.

Not only must the enclosures be technically appropriate but the supplier must provide the proper certifications regarding its approval by an organization such as Factory Mutual. It is then the purchaser's responsibility to make sure that the classification for the enclosure is acceptable for the hazard area classification of the location in which the measurement device and enclosure will be installed. All of this information must be retained; electrical inspectors like to look at these things.

Purged Enclosures

Flammable gases and vapors are kept away from the electrical equipment within the enclosure by pressurizing and purging. NFPA standard 496, *Purged and Pressurized Enclosures for Electrical Equipment*,[13] provides for three types:

X. Reduces classification from Division 1 to nonhazardous.

Y. Reduces classification from Division 1 to Division 2.

Z. Reduces classification from Division 2 to nonhazardous.

For type X, power is to be disconnected from the internal electrical components on loss of purge, as shown in Figure 1.35. The protective switch must be installed in the discharge line, and no valve or other restriction can be installed between the protective switch and the enclosure. Refer to the standard for additional requirements on purged enclosures.

In Europe, the requirements for pressurized enclosures are incorporated into IEC 60079 as *Part 2—Pressurized Enclosures*.[14] Actually, *pressurized* is a more appropriate term than *purged*—the instrumentation actually ensures that the pressure within the enclosure is above atmospheric, which means that any leaks are from inside to outside. The flow through the enclosure often depends on how well it is sealed.

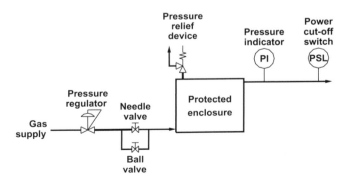

Figure 1.35. Pressurized enclosure.

Intrinsic Safety

For many applications, electrical circuits can be designed so as to be incapable of providing a spark or other source of ignition on circuit or component failures. Referred to as *intrinsically safe*, such designs eliminate the need for explosion-proof or purged enclosures.

Electrical devices are intrinsically safe only if certified as such. These certifications take two forms:

System. Elements of the circuit are approved as a system. The manufacturer submits all components of the circuit that contains the measurement device for approval. The measurement device is certified as intrinsically safe only when used in that circuit.

Entity. Individual components are approved for use in properly designed systems. This gives the user more flexibility but also imposes more responsibility on the user to get it right.

Intrinsically safe designs are possible only in low-power applications (which covers most but not all measurement devices). But low power alone does not mean intrinsically safe. The circuits must not be capable of storing sufficient energy, either via capacitance or inductance, that could lead to a spark.

To summarize, the advantages and disadvantages of intrinsically safe electronics are as follows:

- Expensive enclosures are not mandatory.
- Restricted to low-power applications.
- Maintenance can be performed without shutting off power.
- Imposes demands on engineering, especially when entity certifications are used.
- Must be installed and maintained by knowledgeable personnel.
- Excellent acceptance internationally.

Intrinsic Safety Barriers

Intrinsic safety barriers limit the current, voltage, and total energy delivered to a measurement device in a hazardous area. Two types are commercially available:

Galvanic isolators. These use transformers, optical isolators, or similar technology to meet the requirements for intrinsic safety. These require an external source of power (active devices).

Zener barriers. These use Zener diodes, resistors, and fuses to limit current and voltage. These do not require an external source of power (passive devices).

Most industrial installations use Zener barriers, and only these are described herein. There are several commercial suppliers of Zener intrinsic safety barriers.

The installation of the intrinsic safety barriers is always in the nonhazardous area. There are two possibilities:

Between the input/output (I/O) equipment and the measurement device. The measurement device must be intrinsically safe, but the I/O equipment does not (Fig. 1.36a). To date, most industrial installations have been in this fashion.

Between the communications/power supplies and the I/O equipment. The I/O equipment and all measurement devices must be intrinsically safe (Figure 1.36b). Being the leaders in the use of intrinsically safe installations, the European suppliers were the first to offer intrinsically safe I/O equipment that could be installed and serviced in the hazardous area.

Figure 1.37 illustrates the circuit for a simple Zener intrinsic safety barrier for positive polarity. The negative side of the power supply is grounded, so the

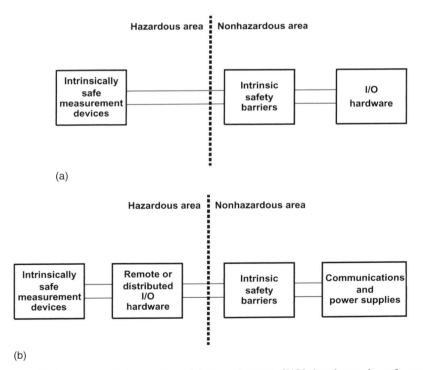

Figure 1.36. Intrinsic safety barriers. (a) Input/output (I/O) hardware in safe area. (b) I/O hardware in hazardous area.

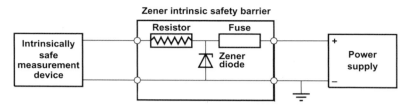

Figure 1.37. Single-channel intrinsic safety barrier, positive polarity.

safety barrier will experience only positive polarity. Circuits are available for negative polarity and alternating polarity. Single-channel means one side is grounded; dual-channel means that neither side is grounded. Consequently, the suppliers offer a large number of designs, with the selection also depending on

- Maximum voltage in safe are.
- Maximum current in safe area.
- Maximum power in safe area.
- Maximum permissible inductance in safe area.
- Maximum permissible capacitance in safe area.

Safety barriers are available for most signals, including

- Two-wire transmitters (4 to 20 ma).
- Thermocouples.
- RTDs (including three wire).
- Process switches (pressure, level, etc.).
- Outputs (to IP converters; 4 to 20 ma).
- Solenoid valves.
- Communication links.

Most suppliers provide a mounting rail or other mechanism that also provides an electrical ground to the barriers.

Attention must be paid to the details. The measurement device must be approved for use in the hazardous area in conjunction with an approved intrinsic safety barrier appropriate for that measurement device. The electrical inspectors start with the measurement device and then proceed to the safety barrier. They expect to see the proper documentation and only equipment with the appropriate certifications for the classification of the hazardous area.

1.19. MEASUREMENT DEVICE DYNAMICS

If the variable being sensed changes, how rapidly does the measurement device output respond to these changes? The answer to this question pertains to the dynamic characteristics of the measurement device. Measurement device dynamics arise more frequently in applications where the output of the measurement device is the input to a process control function.

Temperature Probe

In Figure 1.38, the fluid temperature T_F is measured using a bare temperature probe. Even for this simple configuration, the equations are relatively simple, provided certain assumptions are made. Finite element analysis are required for a more accurate analysis and can encompass the thermowell that is usually present.

Let's use the following notation:

A	Heat transfer area, m^2 or ft^2
c_P	Specific heat of probe, kcal/kg $\cdot$ °C or Btu/lb$_m \cdot$ °F
M	Mass of probe, kg or lb$_m$
T_F	Fluid temperature, °C or °F
T_P	Probe temperature, °C or °F
T_{ref}	Reference temperature, °C or °F
U	Heat transfer coefficient, kcal/min $\cdot$ m$^2 \cdot$ °C or Btu/min $\cdot$ ft$^2 \cdot$ °F

Let there be no steady-state error—that is, $T_P = T_F$ at equilibrium. Suppose the fluid temperature T_F increases abruptly by 5 °C. How does the probe tem-

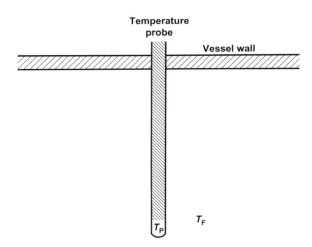

Figure 1.38. Bare probe for temperature measurement.

perature T_P respond? The probe temperature will not increase as rapidly as the fluid temperature. How rapidly the probe responds depends on two factors:

> *Heat transfer coefficient.* The greater the heat transfer coefficient, the greater the heat transfer for a given temperature difference. Because this improves the response of the probe, the probe should be installed at a location where the fluid flows rapidly across the probe, which increases the heat transfer coefficient.

> *Mass of the probe.* The greater the mass of the probe, the more heat that must be transferred to increase the temperature of the probe by $1\,°C$. For rapid response, the mass of the probe should be as small as possible.

Time Constant

An energy balance around the probe involves the following two terms:

Heat transfer rate $\qquad\qquad UA(T_F - T_P)$
Energy content of the probe $\qquad Mc_P(T_P - T_{\text{ref}})$

The rate of change of energy within the probe is equal to the heat transfer rate:

$$\frac{d}{dt}[Mc_P(T_P - T_{\text{ref}})] = Mc_P\frac{dT_P}{dt} = UA(T_F - T_P)$$

$$\frac{Mc_P}{UA}\frac{dT_P}{dt} + T_P = T_F$$

The coefficient Mc_P/UA has units of time:

Metric $\qquad\qquad \dfrac{Mc_P}{UA} = \dfrac{(\text{kg})(\text{kcal/kg}\cdot°C)}{(\text{kcal/min}\cdot\text{m}^2\cdot°C)(\text{m}^2)} = \text{min}$

English $\qquad\qquad \dfrac{Mc_P}{UA} = \dfrac{(\text{lb}_m)(\text{Btu/lb}_m\cdot°F)}{(\text{Btu/min}\cdot\text{ft}^2\cdot°F)(\text{ft}^2)} = \text{min}$

This coefficient is called the *time constant* of the probe. The value of the time constant determines how quickly the probe responds. A probe with a small mass M and a high heat transfer coefficient U responds quickly (value of Mc_P/UA is small) and is said to have a *short* time constant. Conversely, a probe with a large mass M and a low heat transfer coefficient U responds slowly (value of Mc_P/UA is large) and is said to have a *long* time constant.

First-Order Lag

Using the Greek letter τ for the time constant, the differential equation for the probe is written as follows:

$$\tau \frac{dT_P}{dt} + T_P = T_F$$

This is a first-order differential equation. Such processes are referred to as first-order lags.

Suppose the fluid temperature abruptly changes from 60° to 65 °C (a step change of +5 °C). The response of the probe temperature is presented in Figure 1.39 for τ = 1.0 min. When the step change is introduced, the probe temperature immediately begins to respond. But as the probe temperature approaches the fluid temperature, the rate of change becomes progressively slower in an exponentially decaying manner.

For the response illustrated in Figure 1.39, the time constant is 1.0 min. Note that the response attains 63% of the total change in 1.0 min. The 63% point is often used to characterize the speed of response of a process, a measurement device, a valve, etc.

Dead Time or Transportation Lag

In the illustration in Figure 1.40, material is being metered from a feed hopper through a rotary valve onto a belt conveyor. The velocity of the belt conveyor is V ft/min; the distance is L ft.

The time t required to transport material from one end of the conveyor to the other is the length L of the conveyor divided by the velocity V of the

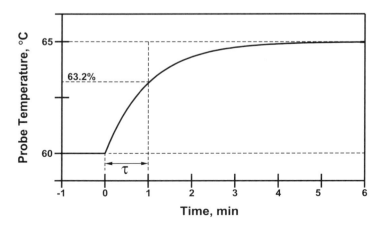

Figure 1.39. Response of a first-order lag to a step change in its input.

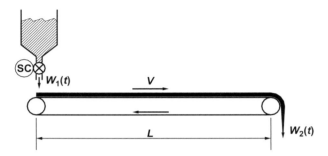

Figure 1.40. Belt conveyor.

conveyor. Although transportation time or transportation lag are more descriptive terms, this time is often referred to as the *dead time* and is designated by θ. For transportation processes,

$$\theta = \frac{L}{V}$$

Material is metered onto the belt conveyor at rate $W_1(t)$. Material falls off the other end of the conveyor at rate $W_2(t)$. The relationship between $W_1(t)$ and $W_2(t)$ is quite simple. The current value for $W_2(t)$ is the value of $W_1(t)$ at 1 dead time in the past. Mathematically, this is expressed as follows:

$$W_2(t) = W_1(t - \theta)$$

The conveyor in Figure 1.40 is a material transport system. Fluid flowing through pipes is another material transport system. Any material transport system exhibits dead time for parameters such as temperature and composition.

Measurement Device Location

When measuring temperatures, dead time can be introduced through the location of the measurement device. The process illustrated in Figure 1.41 mixes steam and cold water to produce hot water. The hot water temperature transmitter has been installed at a distance L from the hot water tank.

A belt conveyor is a material transport system; so is fluid flowing through pipes. The time required for the water to flow from the tank to the transmitter location is the distance L divided by the velocity V of the water flowing through the pipe. Locating the transmitter downstream of the tank introduces dead time:

Transmitter location introduces dead time for measurements of an intensive property (temperature, composition, physical properties, etc). But for measurements of an extensive property (flow, level, etc), sensor location does not

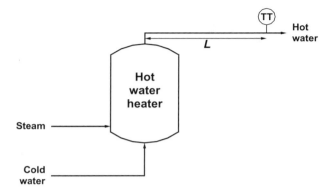

Figure 1.41. Introducing dead time through the location of the temperature transmitter.

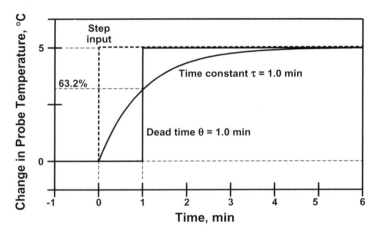

Figure 1.42. Responses to a step change in the input.

introduce dead time. That is, a flow transmitter located a distance L from the tank would indicate the same flow as one located at the tank.

Dead time is always associated with the transport of material from one location to another. This becomes a major concern for analyzer sample systems such as the one shown in Figure 1.6.

Step Response

A customary practice is to present responses as the change in the variable of interest from its starting or initial values. When presented in this manner, the responses always start at zero.

The responses to a step increase of $5\,°C$ for two systems are shown in Figure 1.42:

- A system consisting of a single time constant, with $\tau = 1.0$ min.
- A system consisting of a pure dead time, with $\theta = 1.0$ min.

The time constant system reacts immediately, and in one time constant attains 63% of the total change. The dead time system exhibits no reaction at all until the dead time has elapsed, and then the change in the input immediately appears on the output.

For each, the 63% point corresponds to the lag in the system. However, it does not describe the nature of the lag. For the time constant system, it is a first-order lag; for the dead time system, it is a transportation lag.

When the measured variable is the input to a controller, dead time is far more detrimental to performance than a time constant.

Ramp Response

Figure 1.43 presents the responses to a ramp increase of $5\,°C/$min for two systems:

- A system consisting of a single time constant, with $\tau = 1.0$ min.
- A system consisting of a pure dead time, with $\theta = 1.0$ min.

The time constant system reacts immediately, but after some time (about 3 min, or 3τ) the response is a ramp that is 1 min later than the ramp input. The dead time system exhibits no reaction at all until the dead time has elapsed, and then the ramp immediately appears on the output. But after approximately 3 min, both outputs are ramps that are 1 min later than the ramp input. The time difference between the ramps is the lag in the system. For the

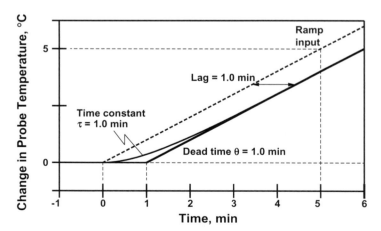

Figure 1.43. Responses to a ramp change in the input.

time constant system, this lag equals the value of the time constant. For the dead time system, this lag equals the value of the dead time.

Dynamic testing involves obtaining the response of the system to an input. We have exhibited this for step inputs and ramp inputs. Another form of the test is the pulse or "bump" test where the input is changed (increased or decreased) for some period of time and then returned to the original value.

Dynamic Performance Measures

When seeking data on the dynamic performance of a measurement device, expect to be disappointed. First, very little information is usually available on the dynamic performance of the measurement device. Second, whatever data are available are often based on applying the measurement device to air or water.

There is no uniform basis for expressing dynamic performance. Some state the time to attain 63% of the total response. This is the lag, but is it a first-order lag or a transportation lag? Most measurement devices behave more like a first-order lag, but a few (notably analyzers) exhibit dead time. Then you will encounter statements such as "90% of the response is attained in less than 20 sec." How do you interpret this?

The common denominator for comparing the dynamic performance of measurement devices has to be the lag. The relationship among the time t, the time constant τ, and the fraction response $c(t)$ at time t is as follows.

$$c(t) = 1 - e^{-t/\tau}$$

or

$$\tau = \frac{-t}{\ln[1 - c(t)]}$$

The values of $c(t)$ at integer multiples of the time constant τ are as follows:

Time	Response, %
τ	63.2
2τ	86.5
3τ	95.0
4τ	98.2
5τ	99.3
6τ	99.8

What is the lag for a measurement device that attains 90% of the change in 20 sec? The list does not contain an entry for 90%, so the lag must be computed from the formula:

$$\tau = \frac{-t}{\ln[1-c(t)]} = \frac{-20}{\ln[1-0.9]} = 8.7 \sec$$

1.20. FILTERING AND SMOOTHING

The objective of any measurement is to provide a signal that indicates the current value of the process variable of interest. However, this is almost always complicated by the presence of some combination of the following:

Noise of process origin. There may be situations within the process that lead to variability in the variable of interest. Here are a couple of examples:
- Radar level transmitters can detect ripples on the surface caused by agitation within a vessel.
- Coriolis flowmeters respond rapidly enough to sense the pulsating flow from a positive displacement pump.

Measurement noise. This is especially prevalent in some measurements, such as weight gauges. However, it can also occur in flow, level, and pressure. When a thermowell is present, noise in temperature measurements is unusual and is likely the result of electrical problems. However, measurement noise is common on temperatures measured by pyrometers.

Stray electrical pickup. Every analog signal contains stray electrical pickup, the source being AC power at either 50 or 60 Hz. For current loop inputs, the analog input processing hardware in digital systems is specifically designed to remove this component.

Options for Smoothing

There are three possibilities for providing filtering or smoothing:

Between the process and the measurement device. This not recommended! However, it is done, often without official knowledge or sanction. For example, liquid level measurements are often installed in an external chamber that is connected to the process vessel at the bottom (liquid connection) and at the top (gas or vapor connection). Smoothing, usually to an unknown but excessive degree, is achieved by partially closing the isolation valve in the liquid (bottom) connection.

Within the measurement device. Most conventional transmitters provide a "damping" setting, usually via an uncalibrated screw adjustment. With no calibration, zero or no smoothing is the only setting for which the amount of smoothing is known. Smart transmitters also provide smoothing, but with the smoothing coefficient specified via a configuration parameter.

Within the data acquisition software. The input processing routines for digital systems provide the option for applying smoothing to the input. Before the advent of smart transmitters, this was the preferred approach.

The important aspect is to know exactly how much smoothing is being provided. For smart transmitters, providing smoothing within the transmitter is a viable alternative and is usually the preferred approach.

Frequency

A signal such as the output of a measurement device can be expressed as the sum of sinusoidal signals of various frequencies and amplitudes. The amplitude of the sinusoidal component at a given frequency can be computed by the Fourier integral:

$$G(j\omega) = \int_0^\infty x(t)e^{-j\omega t}dt$$

where $x(t)$ = signal; ω = frequency (radians/sec); t = time (sec).

The amplitude at frequency ω is $|G(j\omega)|$. A plot of $|G(j\omega)|$ as a function of frequency ω is a good way to characterize the nature of a signal. Unfortunately, such plots are rarely generated for the outputs from measurement devices. To obtain useful data from the Fourier integral, high scan rates (such as 100 samples per second) are normally required. The data acquisition equipment customarily installed in process applications cannot achieve such rates.

In process applications, the content of interest is usually at the lower frequencies. Noise of process origin is normally at a higher frequency, measurement noise is at an even higher frequency, and stray electrical pickup (at 50 or 60 Hz) is usually the highest component of significance.

The units for frequency in the Fourier integral are radians per second. The more common units are cycles per second, or Hertz (Hz). There are 2π radians per cycle, so multiply the frequency in Hertz by 2π to obtain the frequency in radians per second.

Filter

Smoothing is provided by a filter. In block diagrams, the filter is normally inserted between the measurement device and the digital system. Figure 1.44 illustrates a control loop that contains a filter. For analog equipment, the filter is a physical piece of hardware usually attached to the input terminals. For digital systems, the filter is provided in the software. With conventional (dumb) transmitters, the filter is normally incorporated into the data acquisition software. With smart transmitters, it is usually preferable to use the filter incorporated into the measurement device's software.

In process applications, the filter is a *low-pass filter* that attenuates the high-frequency components of a signal but not the low-frequency components. An

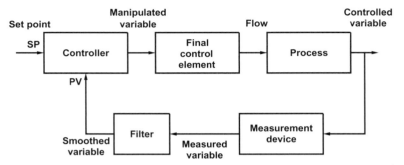

Figure 1.44. Control loop with a filter on the measured value.

ideal low-pass filter completely removes all components above a specified frequency (the cutoff frequency) and passes unaltered all components below this frequency. Unfortunately, such filters are unrealizable.

Practical low-pass filters do not provide a sharp cutoff as in the ideal filter. The cutoff frequency of a practical filter is normally defined as the frequency for which the attenuation factor is 0.891. The attenuation is greater for frequencies above the cutoff frequency (the higher the frequency, the greater the attenuation). Components at frequencies well below the cutoff frequency are passed basically unaltered, but some attenuation occurs for frequencies just below the cutoff frequency.

Exponential Filter

The exponential filter is described by the following differential equation.

$$\tau_F \frac{dy(t)}{dt} + y(t) = x(t)$$

where $x(t)$ = input to filter; $y(t)$ = output from filter; τ_F = filter time constant (sec); t = time (sec).

This equation is exactly the same as the equation for a first-order lag with time constant τ_F. The step response and the ramp response of the exponential filter also exhibit the same behavior as a first-order lag. Therefore, the following have the same effect:

- Applying an exponential filter to the output of a measurement device.
- Inserting a temperature measurement device into a thermowell.

In analog systems, the exponential filter is implemented in hardware, the most common being the resistor-capacitor (RC) network filter. The differential equation just given applies to the RC network filter, with $\tau_F = 1/RC$. In digital systems, the exponential filter is implemented in software using a difference equation that is derived from the first-order lag equation as follows.

Differential equation for exponential filter:

$$\tau_F \frac{dy(t)}{dt} + y(t) = x(t)$$

Difference equation with backward difference to approximate the derivative:

$$\tau_F \frac{y_i - y_{i-1}}{\Delta t} + y_i = x_i$$

Solve difference equation for y_i:

$$y_i = \frac{\Delta t}{\tau_F + \Delta t} x_i + \frac{\tau_F}{\tau_F + \Delta t} y_{i-1}$$

Introduce smoothing coefficient k:

$$y_i = kx_i + (1-k)y_{i-1}$$

$$k = \frac{\Delta t}{\tau_F + \Delta t}$$

where Δt = time between execution (also known as the *sampling time*).

An alternate derivation gives $k = 1 - \exp(-\Delta t / \tau_F)$, but the expression for k derived above is equivalent for $\Delta t \ll \tau_F$. In some systems, the degree of smoothing is specified as the smoothing coefficient k (note that $0 < k < 1$) of the difference equation, but the trend is to specify the filter time constant τ_F.

Moving Average Filter

The moving average filter smoothes by computing the arithmetic average of some number N of consecutive input values.

$$y_i = \frac{1}{N} \sum_{j=0}^{N-1} x_{i-j}$$

The smoothing can be specified as either of the following:

- Averaging time T_A, usually in seconds. All input values received during this period of time are averaged.
- Number of input values N to be averaged.

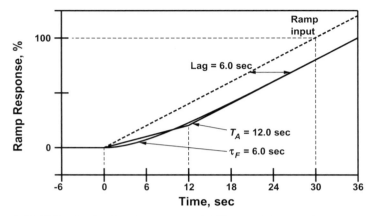

Figure 1.45. Effective lag for the exponential filter and the moving average filter.

Note that $N = T_A/\Delta t$, where Δt is the interval (in seconds) between values.

What is the origin of the term *moving*? The filter maintains a storage array for the required number N of input values. When a new input value is received, the new value replaces the oldest value in the storage array, and the average is recomputed. Unlike the exponential filter, there is no practical analog equivalent to the moving average filter.

Although not identical, the performance of the two filters is very similar when $T_A = 2\tau_F$. Figure 1.45 presents the ramp responses for an exponential filer with $\tau_F = 6$ sec and a moving average filter with $T_A = 12$ sec. The ramp response for the moving average filter clearly illustrates that the lag is $T_A/2$.

Use and Abuse of Filters

Specifying a value for the smoothing coefficient (τ_F for the exponential filter; T_A for the moving average filter) determines the cutoff frequency for the filter. To do so precisely, the first step is to characterize the input signal in terms of its components at various frequencies. In the process industries, this is rarely done, one reason being that it requires a scan rate higher than can be achieved by most industrial data acquisition systems.

In the process industries, the usual approach for setting the smoothing coefficient is to observe the variations in the measured variable and then increase the smoothing until the variations are insignificant. The typical result is excessive smoothing.

To make matters worse, filtering is often used to hide or obscure problems, either within the process, the measurement device, or the wiring. The presence of noise in the signal from a temperature measurement device inserted into a thermowell is most likely due to wiring problems. But instead of locating the wiring problem and correcting it, the temptation to hide the problem with smoothing is apparently irresistible. Coriolis flow meters respond rapidly enough to detect the pulsating flow from a positive displacement pump. A pulsation damper should be installed, properly pressurized, and otherwise in working order. Smoothing is not an alternative to a properly functioning pulsation damper.

LITERATURE CITED

1. International Organization for Standardization, *International vocabulary of metrology—Basic and general concepts and associated terms*, ISO/IEC Guide 99, ISO, Geneva, 2007.

2. International Organization for Standardization, *Guide to the Expression of Uncertainty in Measurement*, ISO/IEC Guide 98, ISO, Geneva, 1995.

3. International Organization for Standardization, *Measurement of Fluid Flow by Means of Pressure Differential Devices Inserted in Circular Cross-Section Conduits Running Full—Part 2: Orifice Plates*, ISO 5167-2, ISO, Geneva, 2003.

4. Adams, Thomas M., *A2LA Guide for the Estimation of Measurement Uncertainty in Testing*, G-104, American Association for Laboratory Accreditation, Frederick, MD, July 2002.

5. International Organization for Standardization, *General Requirements for the Competence of Testing and Calibration Laboratories*, ISO/IEC 17025, ISO, Geneva, 2005.

6. *Modicon Modbus Protocol Reference Guide*, PI–MBUS–300, Rev. J, Modicon, Int. Andover, MA, June 1996.

7. International Electrotechnical Commission, *Industrial Communications Networks—Fieldbus Specifications*, IEC 61158 (all parts), IEC, Geneva, 2007.

8. National Electrical Manufacturers Association, *Enclosures for Electrical Equipment (1000 Volts Maximum)*, NEMA Standards Publication 250-2003, NEMA, Rosslyn, VA, 2003.

9. International Electrotechnical Commission, *Degrees of Protection Provided by Enclosures (IP Code)*, IEC 60529, IEC, Geneva, 2001.

10. R. Stahl/AG, *Basics of Explosion Protection*, available at www.rstahl.com.

11. National Fire Protection Association, *National Electrical Code*, NFPA 70, NFPA, Quincy, MA, 2008.

12. International Electrotechnical Commission, *Electrical Apparatus for Explosive Gas Atmospheres—Part 10: Classification of Hazardous Areas*, IEC 60079-10, IEC, Geneva, 2002.

13. National Fire Protection Association, *Standard for Purged and Pressurized Enclosures for Electrical Equipment*, NFPA 496, NFPA, Quincy, MA, 2008.

14. International Electrotechnical Commission, *Explosive Atmospheres—Part 2: Equipment Protection by Pressurized Enclosures*, IEC 60079-2, IEC, Geneva, 2007.

CHAPTER 2

Temperature

Temperature is an extremely important process variable:

- All chemical reactions are temperature sensitive, in terms of both rate and product distribution.
- Most physical properties (density, viscosity, etc.) are influenced by temperature.
- Vapor–liquid equilibria relationships are affected by temperature.
- Plant equipment must be operated in accordance with its temperature specifications.

The importance of temperature measurement in the process industries is reason enough to begin with temperature. We first review the concept of temperature and temperature scales. We then discuss thermowells, which are usually required for measurements of fluid temperatures. We then discuss various measurement technologies, including thermocouples, resistance temperature detectors (RTDs), thermistors, and pyrometers.

2.1. HEAT AND TEMPERATURE

We commonly use terms such as *hot*, *warm*, *cool*, and *cold*. Then we make comparisons, such as, "It is hotter today than yesterday". Temperature is a measure that quantifies such terms.

The temperature of a body is determined by the kinetic energy of the molecules within the body. However, temperature is not a measure of energy but, instead, is a measure of the ability of an object to transfer heat to another object.

When we say that two objects are at the same temperature, this does not mean that the molecules within the two objects have the same kinetic energy. It means that should the two objects come into contact, no heat will flow from one object to the other.

Basic Process Measurements, by Cecil L. Smith
Copyright © 2009 by John Wiley & Sons, Inc.

Heat Transfer

Place a copper object in direct contact with a steel object. There are three possible results:

- *Heat flows from the copper object to the steel object.* In relative terms, the copper object is "hot" and the steel object is "cold." The temperature of the copper object is higher than the temperature of the steel object.
- *Heat flows from the steel object to the copper object.* In relative terms, the steel object is hot and the copper object is cold. The temperature of the steel object is higher than the temperature of the copper object.
- *No heat flow occurs.* The objects are said to be at thermal equilibrium. The objects are at the same temperature.

Heat is said to flow *downhill*—that is, heat flows from a higher temperature to a lower temperature.

Thermal Equilibrium

Two objects are in thermal equilibrium when no heat flows from one object to another. The two objects have the same temperature.

Thermal equilibrium is important in the context of a measurement device. For devices that must make contact with the object whose temperature is to be measured, the sensor, or transducer part, of the measurement device must be in thermal equilibrium with the object. Only then will the temperature reported by the measurement device be the same as the temperature of the object of interest.

Of all the temperature measurement devices that will be discussed subsequently, only the pyrometer is a noncontact temperature measurement device.

Absolute Zero

Absolute zero is the temperature at which all molecular motion stops (molecules have no kinetic energy). A value has been established for the temperature at absolute zero, but this temperature can never be attained. This is the basis for the Third Law of Thermodynamics. It is not possible for an object to have a temperature lower than absolute zero.

Thermometer

A thermometer is a temperature measuring system that encompasses both a sensor and an indicator.

Traditionally, an individual's first contact with a thermometer is when one of the liquid-in-glass variety was inserted into a bodily orifice, one of which tends to be quite uncomfortable. However, these have largely been replaced by digital thermometers, which consist of a sensor coupled with a digital display. These are sometimes still inserted into uncomfortable places, but at least they respond more rapidly. They also produce an audible sound when thermal equilibrium has been attained (actually, when the temperature sensed by the thermometer ceases to change).

2.2. TEMPERATURE SCALES

Over the years, a surprisingly large number of temperature scales have appeared. The first quantitative temperature scale, *the Rømer scale*, was developed in 1692 by the Danish astronomer Olef Rømer. The French scientist René Antoine Ferchault de Réaumur developed the *Réaumur scale*, which was similar to the *Celsius scale*, except that the boiling point of water was 80° on the Réaumur scale. This scale was widely used in Europe before being replaced by the Celsius scale, the only difference being that the boiling point of water is 100 °C. Many of the early efforts centered around what thermometric fluid to use (Isaac Newton used linseed oil as the thermometric fluid).

Today only the Celsius and *Fahrenheit* scales are encountered in everyday use; the absolute scales *Kelvin* and *Rankine* are also used within scientific circles. The equations for converting a temperature from one scale to another are as follows:

Scale	Conversion Equations
Celsius (°C)	$T(°C) = (T(°F) - 32)/1.8$
	$T(°C) = T(K) - 273.15$
	$T(°C) = T(°R)/1.8 - 273.15$
Fahrenheit (°F)	$T(°F) = T(°C) \times 1.8 + 32$
	$T(°F) = T(K) \times 1.8 - 459.67$
	$T(°F) = T(°R) - 459.67$
Kelvin (K)	$T(K) = T(°C) + 273.15$
	$T(K) = (T(°F) + 459.67)/1.8$
	$T(K) = T(°R)/1.8$
Rankine (°R)	$T(°R) = (T(°C) + 273.15) \times 1.8$
	$T(°R) = T(°F) + 459.67$
	$T(°R) = T(K) \times 1.8$

Celsius Temperature Scale

The Celsius (after Anders Celsius, a Swedish astronomer) temperature scale is based on the following two points (both at atmospheric pressure):

0 °C, the freezing point of water.
100 °C, the boiling point of water.

On this scale, absolute zero is −273.15 °C.

The Celsius temperature scale is sometimes referred to as the centigrade temperature scale. The Celsius temperature scale is used almost universally outside the United States.

Fahrenheit Temperature Scale

Gabriel Daniel Fahrenheit was a scientific instrument manufacturer who devised the Fahrenheit temperature scale in 1714. The lowest temperature he could obtain using a mixture of ice and salt was assigned the value of 0 °F (Fahrenheit's expertise was glass; lower temperatures can be obtained using mixtures of ice and salt). He then assigned 96 °F to be his wife's temperature. It is amazing that such a scale would eventually be used on a widespread basis. But in addition to being good with glass, Fahrenheit recognized the advantages of using mercury over alcohol in his thermometers.

The Fahrenheit temperature scale is now based on the following two points (both at atmospheric pressure):

32 °F, the freezing point of water.
212 °F, the boiling point of water.

On this scale, absolute zero is −459.67 °F.

Absolute Temperature Scales

The Kelvin temperature scale is the Celsius temperature scale biased so that absolute zero is 0 K. To convert from Celsius to Kelvin, simply add 273.15.

The Rankine temperature scale is the Fahrenheit temperature scale biased so that absolute zero is 0 °R. To convert from Fahrenheit to Kelvin, simply add 459.67.

The Kelvin and Rankine temperature scales differ by a factor of 1.8. To convert from Kelvin to Rankine, simply multiply by 1.8.

$0 K$ is not a typo. The freezing point of water is 0°C, 32 °F, 491.67 °R, or 273.15 K. Although this seems inconsistent, technically the degree symbol ° should be omitted when using the Kelvin scale. But do not be surprised if you see the freezing point of water stated as 273.15 °K.

Triple Point

For any substance, the liquid phase, solid phase, and vapor phase can coexist at only one temperature and pressure. This is known as the *triple point*.

Figure 2.1 presents the phase diagram for water. At the triple point, the temperature is 0.01 °C and the pressure is 0.006032 atm (0.08647 psia). The melting point is only slightly influenced by pressure, but the boiling point increases with pressure.

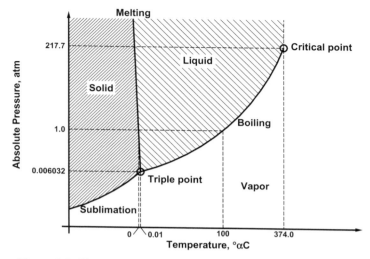

Figure 2.1. Phase diagram for water illustrating the triple point.

Start with ice at atmospheric pressure (1 atm), which is above the triple point. At 0 °C, the ice melts to liquid water. At 100 °C, the liquid water vaporizes. There is a transition from solid to liquid to vapor.

At pressures below the triple point, water cannot exist in the liquid state—that is, at pressures below 0.006032 atm, ice changes directly into vapor, which is known as sublimation. The triple point for carbon dioxide is −56.6 °C and 5.11 atm (75.1 psia). At atmospheric pressure, solid carbon dioxide changes directly into carbon dioxide vapor, hence the name *dry ice* for solid carbon dioxide. At pressures above 5.11 atm, carbon dioxide will make the solid to liquid to gas transitions. Liquid carbon dioxide can exist only at pressures above 5.11 atm.

Thermodynamic Temperature Scale

Technically known as the *Kelvin thermodynamic temperature scale*, this scale is defined by only two points:

0 K, absolute zero.
273.16 K, the triple point of water.

Triple points are ideally suited as the basis for international standards for temperature scales. As stated earlier, for a given substance, there is one and only one temperature at which the three phases (solid, liquid, and vapor) can coexist.

Special equipment is required to replicate a triple point. However, it is reasonable for calibration labs to possess such equipment. Calibration labs that provide calibration services for temperature measurement devices will state what triple points they are able to replicate.

ITS-90

The international temperature scale of 1990 (ITS-90) comprises a number of ranges and subranges, some of which overlap. In calibrating temperature measurement devices, calibration labs rely on the specifications of ITS-90 to perform the calibrations.

Some of the ranges and subranges are defined based on the triple points of various substances, including hydrogen (13.8033 K), neon (24.5561 K), oxygen (54.3584), argon (83.8058 K), mercury (234.3156 K), and water (273.16 K). Below 13.8033 K, the temperature scale is based on the vapor pressure relationships for helium.

Above the triple point of water, the temperature scale is based on the freezing points of metals, including tin (231.928 °C), zinc (419.527 °C), aluminum (660.323 °C), and silver (961.78 °C). Above the melting point of silver, the temperature scale is based on the Plank Radiation Law.

Temperature Difference

The difference between the boiling point of water and the freezing point of water is as follows:

Celsius. $\Delta T(°C) = 100\,°C - 0\,°C = 100\,°C$
Fahrenheit. $\Delta T(°F) = 212\,°F - 32\,°F = 180\,°F$

Being temperature differences, these differ by a factor of 1.8. Temperature differences cannot be converted using the equations for temperatures.

When converting, it is important to distinguish between temperature and temperature difference:

Term	Celsius to Fahrenheit	Fahrenheit to Celsius
Temperature	$T(°F) = 1.8 \times T(°C) + 32$	$T(°C) = [T(°F) - 32]/1.8$
Temperature difference	$\Delta T(°F) = 1.8 \times \Delta T(°C)$	$\Delta T(°C) = \Delta T(°F)/1.8$

The confusing aspect of this is the use of °C and °F for both temperature and temperature difference. One suggestion is to use °C (degrees Celsius) and °F (degrees Fahrenheit) for temperature, but use C° (Celsius degrees) and F° (Fahrenheit degrees) for temperature difference. However, this is not common practice.

2.3. THERMOWELLS

The technologies that are potentially applicable to industrial temperature measurement include the following:

- Liquid in glass.
- Bimetallic strip.
- Filled bulb.
- Thermocouple.
- Resistance temperature detector (RTD).
- Thermistor (nonmetallic RTD).
- Pyrometer.

All of these except the pyrometer require contact with the object whose temperature is to be sensed. The sensor or transducer of all contact temperature measurement devices must be in thermal equilibrium with the substance or object whose temperature is to be sensed. Being a noncontact temperature measurement device, the sensor or transducer of the pyrometer need not be in thermal equilibrium with the substance or object whose temperature is to be sensed.

Thermowell

For a contact temperature measurement device, the fastest response is obtained when the sensor or transducer is installed such that it is in direct contact with the substance or object whose temperature is to be measured. Two factors often make this approach unrealistic for industrial temperature measurements:

- The nature of the materials within the process is such that some adverse effect occurs when there is direct contact with the sensor or transducer.
- For maintenance purposes, access to the sensor or transducer must be possible without disrupting process operations.

For these reasons, the usual practice is to install the sensor or transducer part of a temperature measurement device within a thermowell. The thermowell is basically a metal part that protrudes into the fluid with an internal bore into which the temperature sensitive probe is inserted. There are exceptions, but very few. For example, in low-pressure gas ducts that contain nonhazardous gases (such as air ducts), thermocouples or RTDs can be inserted without a thermowell.

Penalty

From a temperature measurement perspective, a thermowell is most appropriately viewed as a necessary evil. The temperature of the sensor or transducer will be the same as the process temperature only under thermal equilibrium conditions. A thermowell adds two potential sources for measurement error:

Heat conduction. Most thermowells are metallic; metals are very good conductors of heat. If the wall temperature is different from the temperature of the fluid, heat conduction either up or down the thermowell results in a static error in the measured value of the temperature.

Dynamic response. The thermowell has a certain capacity to store and release heat. If the temperature of the fluid is rising, some heat must flow to the thermowell to increase its temperature. While this is occurring, the measured temperature will be below the fluid temperature.

In industrial installations, both are difficult to quantify under process operating conditions. They are likely to be more significant in difficult temperature measurement applications, such as heavy-duty thermowells for high-pressure services or glass-coated thermowells for corrosive services.

Sensor–Thermowell Contact

So that the temperature-sensitive element can be easily inserted and withdrawn, the bore diameter must be slightly larger. Unless one of the following steps is taken, this will impede the heat transfer to the sensor or transducer:

Spring-loaded arrangements. The purpose is to press the tip of the temperature-sensitive element against the end of the bore within the thermowell. For RTDs and thermocouples, the sensitive zone is at the tip, so these elements are routinely supplied with a spring-loaded mechanism.

Heat-conducting fluid. Air is a very poor conductor of heat. Filling the thermowell with a heat-conducting liquid provides far superior heat transfer from the thermowell to the temperature-sensitive element. The downsides of this approach should be rather obvious—it's just messy.

Heat transfer grease or paste. The sensitive zone of the temperature-sensitive element is coated with a special grease or paste before it is inserted into the thermowell. Although less messy than the heat-conducting fluid, the grease or paste tends to accumulate each time the temperature-sensitive element is replaced, unless both the element and the inside of the thermowell are cleaned each time the element is reinserted.

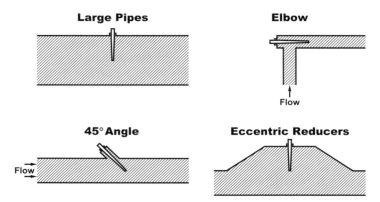

Figure 2.2. Arrangements for installing a thermowell in a pipe.

Installation in Pipes

Figure 2.2 illustrates various arrangements for inserting thermowells to measure the temperature of a fluid flowing in a pipe. When we examine the design of thermowells, we will provide guidelines for determining the immersion depth. In large pipes, the required immersion depth can usually be obtained by inserting the thermowell through the wall of a straight run of pipe.

For small pipes (2 in. or less), other ways are usually necessary to obtain the required immersion depth. Three options are available:

- Insert the thermowell into an elbow of the piping. The direction of flow must be from the base of the thermowell to the tip. With this approach, you can obtain whatever immersion depth is required. In very small pipes, reducers may be required to increase the pipe size so that flow is not restricted by the thermowell.
- Insert the thermowell at a 45° angle. This provides some increase in the available immersion depth, but there are still limits.
- Use eccentric reducers to increase the pipe diameter so that the required immersion depth can be attained. This increases the complexity and costs of the piping.

Installation in Vessels

Figure 2.3 illustrates various arrangements for inserting thermowells to measure the temperature of a fluid in a vessel. The most common installation is to insert the thermowell through a nozzle on the side or bottom of the vessel. This can raise a couple of issues:

Leaks at the nozzle. The seriousness depends on the nature of the chemicals within the vessel, especially if the vessel contents are also under high pressure.

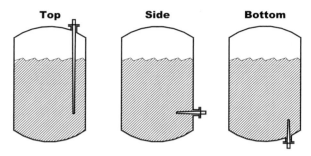

Figure 2.3. Arrangements for installing a thermowell in a vessel.

Buildups at the nozzle. Especially for nozzles at the bottom of a vessel, the nozzle provides a "dead space" whose contents can be quite different from that in the agitated part of the vessel. In some reacting systems, this leads to undesired reactions that produce contaminants that degrade the product quality.

The other alternative is to insert the thermowell through a nozzle at the top of the vessel. The longer the thermowell, the more serious the structural considerations become. In agitated vessels, the thermowell must be sufficiently rigid so that it is not bent by the force of the fluid against the thermowell. These forces tend to be large, and the thermowell becomes quite massive.

The measured temperature is the temperature of the fluid at the physical location of the thermowell. In agitated vessels the temperature is reasonably uniform throughout the vessel. For storage tanks, a pump-around loop provides sufficient agitation that the temperature throughout the vessel is reasonably uniform. But in vessels with no agitation, stratification within the liquid is likely, especially in large storage tanks.

Straight, Tapered, and Stepped

Terms that apply to the shank—the portion of the thermowell that protrudes into the fluid—are *straight*, *tapered*, and *stepped* (Fig. 2.4):

Straight. The shank has a constant outside diameter.

Tapered. The outside diameter of the shank decreases linearly from its maximum value at the base to its smallest value at the tip.

Stepped. The outside diameter of the shank is decreased near the tip. The large diameter at the base provides the necessary structural rigidity. The smaller diameter at the tip reduces the thermal capacity in the region of the sensitive portion of the sensing element, thereby providing a faster response.

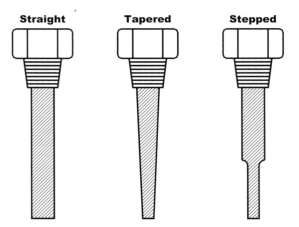

Figure 2.4. Thermowell shank styles.

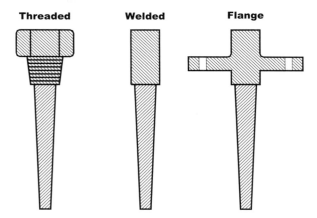

Figure 2.5. Thermowell process connections.

The choice here depends primarily on structural issues. The longer the shank, the more serious these issues become, especially when the flow direction is across the thermowell. We shall examine these issues shortly.

Process Connection

Figure 2.5 illustrates the three options for process connection available in standard products:

Threaded. Sizes of 3/4 in. and 1 in. are most common, but other sizes can be obtained.

Welded. Sizes of 3/4 in. and 1 in. are commonly supplied for socket welding; for welding directly into piping, the size is usually 1½ in.

Flanged. Flanged thermowells are required when inserting a thermowell through a vessel nozzle. The size (1 in. up), rating, face, and material of construction are to the user's specifications.

The shank diameter is normally determined by the size of the process connection element. For example, if the process connection is 3/4-in. normal pipe threads, the shank diameter is 7/8 in. (note that the outside diameter of a 3/4-in. pipe is actually 1.050 in.).

Inside Bore

The inside bore is always straight. Only two inside bore diameters are provided in standard products:

0.260 in. (6.6 mm). This bore is for 1/4-in. elements, including RTDs, #20 gauge thermocouple elements, and 1/4-in. bimetallic thermometers.

0.385 in. (9.8 mm). This bore is for 3/8-in. elements, including #14 gauge thermocouple elements and 3/8-in. bimetallic thermometers.

It should be obvious from this that the dimensions of the thermowell must be consistent with the dimensions of the temperature-sensitive element. For connecting the temperature-sensitive element to the thermowell, the bore is fitted with 1/2-in. pipe threads.

Lengths

Figure 2.6 illustrates various thermowell dimensions. The length of the temperature-sensitive element must be consistent with the following thermowell dimensions:

Insertion (immersion) length. Distance from the base of the process connection (threads, weld, or flange) to the tip of the thermowell.

Lagging extension. Thermowells to be installed in an insulated pipe or vessel require a *lagging extension.* The length of the lagging extension depends on the thickness of the insulation.

Bore length. This is Insertion length + Lagging extension + Process connection (a threaded ¾-in. process connection adds about 1¾ in.) – Thickness of the thermowell at the end of the bore (usually ¼ in. for normal pressure services).

Head-to-thermowell connector. For thermocouples and RTDs, an enclosure called the *head* is required for the wiring connections. The head can be attached directly to the thermowell, but often the head is connected via a pipe and possibly a union coupling.

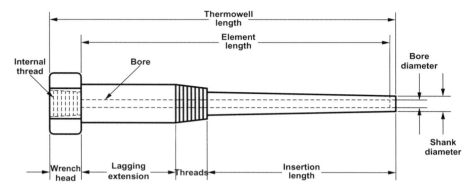

Figure 2.6. Thermowell dimensions.

The length of the temperature-sensitive element is the bore length plus the length of the head-to-thermowell connector.

Insertion Length

The thermowell must extend far enough into the fluid that a representative temperature is obtained. The insertion length is greater for gases than for liquids, the usual guidelines being as follows:

Liquids. Insertion length = 1 in. + length of sensitive zone.
Gases and vapors. Insertion length = 3 in. + length of sensitive zone.

The length of the sensitive zone depends on the nature of the temperature-sensitive element. The sensitive zone of thermocouples and RTDs is basically the point at their tip, so the length of the sensitive zone can be treated as zero. For bimetallic thermometers, the length of the sensitive zone varies with the model, but is usually in the 1- to 2-in. range.

In installations in pipes, the sensitive zone of the temperature-sensing element should be at the centerline of the pipe. However, this location must provide at least the minimum insertion length suggested by the guidelines. Small pipes often do not, an issue that was discussed earlier.

Stress Analysis

The following three stresses arise in thermowell installations:

Static pressure. The pressure rating for the thermowell depends on the thermowell wall thicknesses, material of construction, and temperature. Pressure ratings decrease with temperature, so it is important that the analysis be based on the highest temperature to which the thermowell will be exposed, not the normal operating temperature.

Bending stress due to fluid flow. The drag force exerted by a flowing fluid depends on the fluid velocity and fluid properties (density and viscosity). This issue is most serious for long thermowells in liquid applications.

Flow-induced vibrations. When fluid flows across a thermowell inserted perpendicular to the direction of flow, vibration is induced into the thermowell due to a phenomenon known as the *von Karman effect.*

Computer programs for analyzing these stresses are available from a variety of sources, including thermowell manufacturers, instrument manufacturers, and engineering services companies. For a fee, most suppliers will analyze a specific thermowell application.

Von Karman Effect   When a fluid flows past an object, turbulence in the form of vortices are formed at the object and shed into the flowing stream, alternating from one side to the other. This same effect at a flagpole causes the flag to flutter in the wind. When fluid flows across a thermowell, this turbulence is known as the von Karman wake or von Karman trail. The potential problem is that this turbulence can induce vibrations into the thermowell itself. There are two frequencies of interest:

Von Karman shedding frequency f_s. This is the frequency at which the vortices are formed and shed at the thermowell. The vortex-shedding frequency increases with fluid velocity and decreases with thermowell diameter.

Natural frequency of vibration of the thermowell f_n. This depends on the construction (wall thicknesses, length, etc.) of the thermowell and the modulus of elasticity of the material of construction.

When $f_s = f_n$, the turbulence from the fluid flowing across the thermowell is vibrating the thermowell at its natural frequency. The other stresses usually cause the thermowell to deform, but excessive vibrations cause the thermowell to snap off. Escaping fluid is not the only consequence. The metal part that is now loose in the piping can cause serious damage to valves, pumps, etc.

Maximum Velocity Rating   The customary recommendation is that the von Karman shedding frequency f_s (which increases with velocity) not exceed 80% of the thermowell's natural frequency of vibration f_n (which is a constant for a given thermowell at a given temperature). This permits the maximum allowable flow velocity to be computed for a specific application. These calculations are frequently referred to as the *Murdock calculations.*[1,2]

Vibration is most likely to be a problem in gas flow applications for the following reasons:

- Gas velocities tend to be higher than liquid velocities.
- Thermowells for gas applications require longer insertion lengths. The natural frequency f_n decreases with thermowell length, so problems arise at lower fluid velocities.

Some thermowell manufacturers state the maximum flow velocity for steam and/or water applications for each of their thermowell designs. However, most thermowell manufacturers can perform the calculations for any fluid.

Natural Frequency of Vibration   The thermowell's natural frequency of vibration f_n depends on the following factors:

Physical dimensions of the thermowell. The lower this natural frequency, the lower the maximum flow velocity. The effects are
- The natural frequency decreases with length, in fact, with the square of the length.
- The natural frequency is lower for thin wall thicknesses.

Modulus of elasticity of the material of construction. The lower the modulus of elasticity, the lower the natural frequency.

Anything that decreases the natural frequency will decrease the maximum fluid velocity. Therefore, it is important to determine the minimum value of the natural frequency that will be experienced over the operating range to which the thermowell will be subjected. This turns out to be the maximum temperature, the reasons being

- The modulus of elasticity decreases with increasing temperature.
- The natural frequency decreases with the decreasing modulus of elasticity.

The analysis must reflect the maximum temperature to which the thermowell will be exposed, not the normal operating temperature.

Using a finite element analysis, the natural frequency can be computed for any thermowell design. This is certainly no problem for computer programs but is not something that can be done manually. Although such programs are widely available, we shall present the traditional equations, mainly because they provide better insight into the relationship of the natural frequency to the design parameters for the thermowell.

The natural frequency f_n of the thermowell is given by the following equation:

$$f_n = \frac{K_f}{L^2}\left[\frac{E}{\rho}\right]^{1/2}$$

where f_n = natural frequency; K_f = constant depending on thermowell dimensions; L = insertion length; E = modulus of elasticity of material of construction; ρ = density of material of construction.

The thermowell manufacturer must provide values for the coefficient K_f for each of their thermowell designs (these values are obtained from a finite element analysis). The values are usually presented in one or more tables that permit the following to be taken into consideration:

- Shape of shank (straight, tapered, stepped).
- Diameter of bore.
- Wall thicknesses (standard vs. high-pressure designs).
- Insertion length. The formula suggests that the natural frequency decreases with the square of the insertion length. This is not exactly correct, hence a small dependence of K_f on the insertion length.

Shedding Frequency   The basis for computing the shedding frequency is a dimensionless number known as the *Strouhal number* N_{St}:

$$N_{St} = \frac{f_s D_t}{V}$$

where N_{St} = Strouhal number; f_s = shedding frequency; D_t = thermowell diameter at tip; V = fluid velocity.

The simpler approaches use a constant value of 0.22 for the Strouhal number. But in reality, the Strouhal number is a function of another dimensionless number, the *Reynolds number* N_{Re}:

$$N_{Re} = \frac{DV\rho}{\mu}$$

where N_{Re} = Reynolds number; V = fluid velocity; ρ = fluid density; μ = fluid dynamic viscosity.

The von Karman effect is the basis for the vortex-shedding flow meter. Considerable information is now available on this relationship, and many of the computer programs take into account the effect of the Reynolds number on the Strouhal number.

The computational steps are as follows:

1. Compute the natural frequency f_n.
2. Set the shedding frequency f_s to 0.8 f_n.
3. Compute the maximum velocity V from the shedding frequency.

This procedure suggests that the shedding frequency f_s must be safely below the natural frequency f_n of the thermowell. At least theoretically, it is also possible to operate such that the shedding frequency is safely above the natural frequency. But for obvious reasons, this approach is neither recommended nor popular.

2.4. BIMETALLIC THERMOMETERS

Bimetallic thermometers (also called bimetal thermometers) are based on the following principles:

- Metals expand with increasing temperature. The degree of expansion is described by the coefficient of expansion.
- The coefficient of expansion is not the same for all metals.

In their simplest configuration, a bimetallic thermometer is constructed by bonding two straight metal strips. In Figure 2.7, the metal with the largest coefficient of expansion is on the bottom. As the temperature increases, the lower metal expands more than the upper metal, causing the strip to bend up.

Bimetallic Coils

The deflection of a bimetallic coil is

- Proportional to the square of the length of the strips.
- Proportional to the temperature change.
- Inversely proportional to the thickness of the strips.

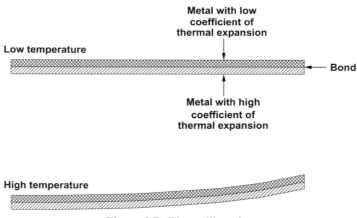

Figure 2.7. Bimetallic strip.

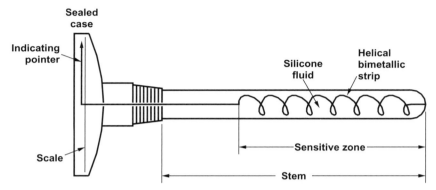

Figure 2.8. Industrial bimetallic thermometer.

Long, thin strips provide the maximum sensitivity. However, this raises serious structural problems for straight strips. There are two alternative arrangements that are more practical:

Spiral. Long lengths can be attained with this arrangement, but the bimetallic strips must be thick enough to support the coil.

Helical. Both long lengths and thin strips are practical. Most industrial bimetallic thermometers use the helix configuration.

Figure 2.8 illustrates the basic components of a bimetallic thermometer with a helical coil. The sensitive zone is usually in the range of 1 to 2 in. Bimetallic thermometers are generally used as indicators only. In catalogs you will occasionally see one with an electrical connection for remote sensing, but these are rarely installed in production facilities.

Configurations

Bimetallic thermometers are available in the following configurations:

- Connection at bottom of indicator dial (Fig. 2.9).
- Connection at back of indicator dial (Fig. 2.9).
- Adjustable configuration by which the indicator can be positioned at any angle between $0°$ and $90°$ of the stem.

The dimensions of off-the-shelf bimetallic thermometers are as follows:

Indicator: 2, 3, or 5 in.
Thermowell connection: 1/2-in. normal pipe threads.
Stem: 1/4 or 3/8 in.; lengths from 2 in. up.

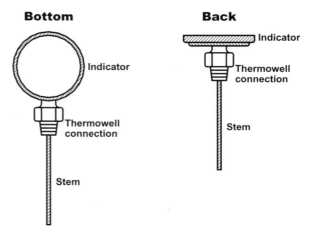

Figure 2.9. Configurations for bimetallic thermometers.

Typical accuracy is ±1% of span. Measurement ranges of −40° to 500°F (−40° to 260°C) are commonly available, with an overrange tolerance of about 50%. Units with an upper range of 1000°F (550°C) are available, but with a lower overrange tolerance. Some bimetallic thermometers are equipped with a recalibration, or reset, adjustment for a single-point calibration.

Temperature Switch

Bimetallic strips are commonly used in temperature switches. In the simplest versions, the deformation of the bimetallic strip determines the state of a mechanical contact. The state of this contact can be sensed remotely. All actuate on rising temperature. These simple products are available in a variety of configurations, some providing only a normally open contact, some providing only a normally closed contact, and some providing both.

Before the digital era, the thermostats in most residential heating and cooling systems contained a bimetallic strip as the temperature-sensitive element. Most of these used a spiral coil for the bimetallic strip coupled with a mercury switch.

2.5. THERMOCOUPLES

Thomas J. Seebeck discovered the underlying principle for the thermocouple in 1821. Consider a closed circuit created by using wires of two dissimilar metals, such as shown in Figure 2.10. There will be two junctions. Let the temperature of one junction be T_1 and the other be T_2. The current flow will be as follows:

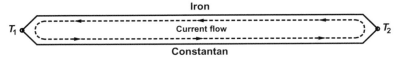

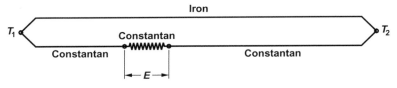

Figure 2.10. The Seebeck effect.

Figure 2.11. Sensing voltage.

$T_1 = T_2$, no current flows.

$T_2 > T_1$, current flow increases with the difference between T_1 and T_2.

$T_2 < T_1$, current flow reverses direction.

The current flow and its direction also depend on the nature of the two dissimilar metals.

The Seebeck effect applies to any two dissimilar metals. But for practical thermocouples, the effect of the temperature difference on current flow must be repeatable, of a reasonable magnitude, and not excessively nonlinear. For these and other reasons (fabrication, resistance to corrosion, cost, etc.), only certain metal pairs are used in industrial thermocouples. This is the basis for thermocouple types. One example is the type J thermocouple in which one metal is iron and the other is constantan (a copper-nickel alloy).

Sensing Voltage

The application of the Seebeck effect to thermocouples involves measuring a voltage, not a current. Suppose one inserts a large resistance at some point in the circuit, as illustrated in Figure 2.11. For the moment, we will fabricate the resistor from the same metal as that of the wire into which it is inserted. For reasons that will be explored shortly, this proves not to be necessary.

Inserting the large resistance reduces the current flow to a very small value. However, this current flow still depends on the difference in the temperatures T_1 and T_2. Because this current is flowing through the resistor, the voltage across the resistor is a function of T_1 and T_2. This voltage is quite low. To give you an idea of the magnitude of the voltage, a type J thermocouple produces just over 5 mV for a temperature difference of 100 °C. Such signal levels in an industrial environment require special (but manageable) wiring practices.

The voltage read-out device used for sensing the thermocouple voltage basically inserts a large resistance into the circuit. This resistance is far larger than the resistance of the wires. This has a beneficial consequence: Changes in the resistance (due to length, temperature, etc.) of the thermocouple wires do not significantly affect the voltage being sensed.

Law of Homogeneous Circuits

In industrial installations, the wiring for a thermocouple may extend over several hundred feet. The temperature along these wires is likely to be near the ambient temperature but will certainly not be uniform. The Law of Homogeneous Circuits basically states that the only temperatures that matter are the temperatures at the junctions.

Figure 2.12 shows that at one point along the upper wire, the temperature is T_3. At a point in the lower wire, the temperature is T_4. The Law of Homogeneous Circuits states that these temperatures affect neither the current flowing around the circuit nor the voltage across the resistor. This definitely makes it easier to apply thermocouples in an industrial environment.

To be completely independent of intermediate temperatures, the thermocouple wires must be homogeneous. This affects the manufacturing process for the thermocouple wires. The purity must be exact and uniform; the wires must be drawn in such a manner that no imperfections are created.

Junctions

In reality, the thermocouple responds to a temperature difference. It cannot be used to directly measure the value of a temperature. But if we know the temperature at one of the junctions, the temperature at the other junction can be computed from the temperature difference. This is the basis of the following terminology:

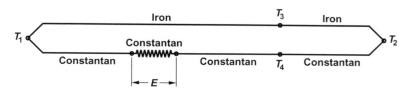

Figure 2.12. Law of homogeneous circuits.

Process junction or measuring junction. This is the junction whose temperature is to be determined.

Reference junction. This is the junction whose temperature is known.

The process, or measuring, junction is commonly referred to as the *hot junction*, and the reference junction as the *cold junction*. But there is one problem with this terminology—in an occasional application the process junction is at a temperature below that of the reference junction.

Law of Intermediate Metals

Let's change the resistor in our circuit to copper, as illustrated by the upper circuit in Figure 2.13. What effect does this have on the current or voltage? None, but with one proviso. Let the temperatures at the terminals of the resistor be T_3 and T_4. As long as these temperatures are the same, using a copper resistor has no effect on the current or voltage.

Inserting a voltage-sensing device into the circuit is equivalent to inserting a copper resistor. Normally, this is inserted at the reference junction, giving the lower circuit in Figure 2.13. We no longer have an iron-constantan junction at the reference junction. Instead, we have an iron-copper junction and a copper-constantan junction. But provided the two terminals at the reference junction are at the same temperature, this is equivalent to an iron-constantan junction.

The two junctions involving copper must be at the same temperature. In practice, this receives special attention, resulting in the so-called isothermal terminal block that contains the termination screws for the thermocouple at the reference junction.

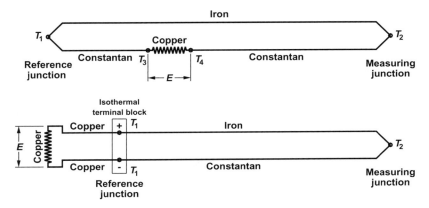

Figure 2.13. Law of intermediate metals.

Voltage–Temperature Relationship

Figure 2.14 presents the voltage as a function of temperature for a type J thermocouple. This graph is for a reference junction temperature of $0\,°C$. One of the advantages of the type J thermocouple is that the graph is nearly linear for temperatures above $0\,°C$. The other types of thermocouples are not so well behaved, but the departure from linearity is generally modest. However, digital systems always provide linearization, even for the type J thermocouple.

The source of the data for preparing the graph is the NIST. For each type of thermocouple, the source data are provided in the form of tables that list the millivolts for various measuring junction temperatures, assuming a reference junction temperature of $0\,°C$. Table 2.1 provides an example of the table for a type J thermocouple; It covers the range of $-210°$ to $200\,°C$; the NIST table covers a range of $-210°$ to $1200\,°C$.

The NIST also provides the coefficients in polynomial equations for computing millivolts from temperature and vice versa.

Compute Millivolts from Temperature   The following polynomial is used for computing millivolts when the temperature is known. The NIST provides the values for the coefficients.

$$E = c_0 + c_1 t + c_2 t^2 + \cdots + c_n t^n$$

where E = voltage (mV); t = temperature ($°C$); n = order of the polynomial.

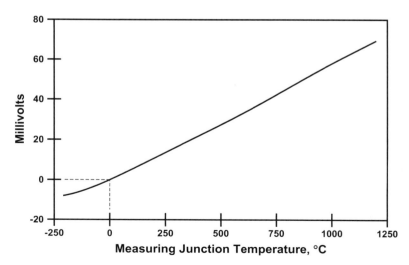

Figure 2.14. Voltage vs. temperature relationship for a type J thermocouple with a reference junction at $0\,°C$.

TABLE 2.1. ITS-90 Table for Type J Thermocouple from −210° to 200°C (Thermoelectric Voltage in Millivolts)

°C	0	−1	−2	−3	−4	−5	−6	−7	−8	−9	−10	°C
−210	−8.095											−210
−200	−7.890	−7.912	−7.934	−7.955	−7.976	−7.996	−8.017	−8.037	−8.057	−8.076	−8.095	−200
−190	−7.659	−7.683	−7.707	−7.731	−7.755	−7.778	−7.801	−7.824	−7.846	−7.868	−7.890	−190
−180	−7.403	−7.429	−7.456	−7.482	−7.508	−7.534	−7.559	−7.585	−7.610	−7.634	−7.659	−180
−170	−7.123	−7.152	−7.181	−7.209	−7.237	−7.265	−7.293	−7.321	−7.348	−7.376	−7.403	−170
−160	−6.821	−6.853	−6.883	−6.914	−6.944	−6.975	−7.005	−7.035	−7.064	−7.094	−7.123	−160
−150	−6.500	−6.533	−6.566	−6.598	−6.631	−6.663	−6.695	−6.727	−6.759	−6.790	−6.821	−150
−140	−6.159	−6.194	−6.229	−6.263	−6.298	−6.332	−6.366	−6.400	−6.433	−6.467	−6.500	−140
−130	−5.801	−5.838	−5.874	−5.910	−5.946	−5.982	−6.018	−6.054	−6.089	−6.124	−6.159	−130
−120	−5.426	−5.465	−5.503	−5.541	−5.578	−5.616	−5.653	−5.690	−5.727	−5.764	−5.801	−120
−110	−5.037	−5.076	−5.116	−5.155	−5.194	−5.233	−5.272	−5.311	−5.350	−5.388	−5.426	−110
−100	−4.633	−4.674	−4.714	−4.755	−4.796	−4.836	−4.877	−4.917	−4.957	−4.997	−5.037	−100
−90	−4.215	−4.257	−4.300	−4.342	−4.384	−4.425	−4.467	−4.509	−4.550	−4.591	−4.633	−90
−80	−3.786	−3.829	−3.872	−3.916	−3.959	−4.002	−4.045	−4.088	−4.130	−4.173	−4.215	−80
−70	−3.344	−3.389	−3.434	−3.478	−3.522	−3.566	−3.610	−3.654	−3.698	−3.742	−3.786	−70
−60	−2.893	−2.938	−2.984	−3.029	−3.075	−3.120	−3.165	−3.210	−3.255	−3.300	−3.344	−60
−50	−2.431	−2.478	−2.524	−2.571	−2.617	−2.663	−2.709	−2.755	−2.801	−2.847	−2.893	−50
−40	−1.961	−2.008	−2.055	−2.103	−2.150	−2.197	−2.244	−2.291	−2.338	−2.385	−2.431	−40
−30	−1.482	−1.530	−1.578	−1.626	−1.674	−1.722	−1.770	−1.818	−1.865	−1.913	−1.961	−30
−20	−0.995	−1.044	−1.093	−1.142	−1.190	−1.239	−1.288	−1.336	−1.385	−1.433	−1.482	−20
−10	−0.501	−0.550	−0.600	−0.650	−0.699	−0.749	−0.798	−0.847	−0.896	−0.946	−0.995	−10
0	0.000	−0.050	−0.101	−0.151	−0.201	−0.251	−0.301	−0.351	−0.401	−0.451	−0.501	0

TABLE 2.1. *Continued*

°C	0	1	2	3	4	5	6	7	8	9	10	°C
0	0.000	0.050	0.101	0.151	0.202	0.253	0.303	0.354	0.405	0.456	0.507	**0**
10	0.507	0.558	0.609	0.660	0.711	0.762	0.814	0.865	0.916	0.968	1.019	**10**
20	1.019	1.071	1.122	1.174	1.226	1.277	1.329	1.381	1.433	1.485	1.537	**20**
30	1.537	1.589	1.641	1.693	1.745	1.797	1.849	1.902	1.954	2.006	2.059	**30**
40	2.059	2.111	2.164	2.216	2.269	2.322	2.374	2.427	2.480	2.532	2.585	**40**
50	2.585	2.638	2.691	2.744	2.797	2.850	2.903	2.956	3.009	3.062	3.116	**50**
60	3.116	3.169	3.222	3.275	3.329	3.382	3.436	3.489	3.543	3.596	3.650	**60**
70	3.650	3.703	3.757	3.810	3.864	3.918	3.971	4.025	4.079	4.133	4.187	**70**
80	4.187	4.240	4.294	4.348	4.402	4.456	4.510	4.564	4.618	4.672	4.726	**80**
90	4.726	4.781	4.835	4.889	4.943	4.997	5.052	5.106	5.160	5.215	5.269	**90**
100	5.269	5.323	5.378	5.432	5.487	5.541	5.595	5.650	5.705	5.759	5.814	**100**
110	5.814	5.868	5.923	5.977	6.032	6.087	6.141	6.196	6.251	6.306	6.360	**110**
120	6.360	6.415	6.470	6.525	6.579	6.634	6.689	6.744	6.799	6.854	6.909	**120**
130	6.909	6.964	7.019	7.074	7.129	7.184	7.239	7.294	7.349	7.404	7.459	**130**
140	7.459	7.514	7.569	7.624	7.679	7.734	7.789	7.844	7.900	7.955	8.010	**140**
150	8.010	8.065	8.120	8.175	8.231	8.286	8.341	8.396	8.452	8.507	8.562	**150**
160	8.562	8.618	8.673	8.728	8.783	8.839	8.894	8.949	9.005	9.060	9.115	**160**
170	9.115	9.171	9.226	9.282	9.337	9.392	9.448	9.503	9.559	9.614	9.669	**170**
180	9.669	9.725	9.780	9.836	9.891	9.947	10.002	10.057	10.113	10.168	10.224	**180**
190	10.224	10.279	10.335	10.390	10.446	10.501	10.557	10.612	10.668	10.723	10.779	**190**
200	10.779	10.834	10.890	10.945	11.001	11.056	11.112	11.167	11.223	11.278	11.334	**200**

112

For the type J thermocouple, there are two temperature ranges:

Range (°C)	Order	Coefficients
−210.000–760.000	8	$c_0 = 0.000000000000 \times 10^{+00}$ $c_1 = 0.503811878150 \times 10^{-01}$ $c_2 = 0.304758369300 \times 10^{-04}$ $c_3 = -0.856810657200 \times 10^{-07}$ $c_4 = 0.132281952950 \times 10^{-09}$ $c_5 = -0.170529583370 \times 10^{-12}$ $c_6 = 0.209480906970 \times 10^{-15}$ $c_7 = -0.125383953360 \times 10^{-18}$ $c_8 = 0.156317256970 \times 10^{-22}$
−760.000–1200.000	5	$c_0 = 0.296456256810 \times 10^{+03}$ $c_1 = -0.149761277860 \times 10^{+01}$ $c_2 = 0.317871039240 \times 10^{-02}$ $c_3 = -0.318476867010 \times 10^{-05}$ $c_4 = 0.157208190040 \times 10^{-08}$ $c_5 = -0.306913690560 \times 10^{-12}$

Compute Temperature from Millivolts   The following equation is used to compute temperature when the millivolts are known. The NIST provides the values for the coefficients.

$$t = d_0 + d_1 E + d_2 E^2 + \cdots + d_n E^n$$

where E = voltage (mV); t = temperature (°C); n = order of the polynomial. For the type J thermocouple, there are three millivolt ranges:

Range (mV)	Order	Coefficients
−8.095–0.000	8	$d_0 = 0.0000000 \times 10^{+00}$ $d_1 = 1.9528268 \times 10^{+01}$ $d_2 = -1.2286185 \times 10^{+00}$ $d_3 = -1.0752178 \times 10^{+00}$ $d_4 = -5.9086933 \times 10^{-01}$ $d_5 = -1.7256713 \times 10^{-01}$ $d_6 = -2.8131513 \times 10^{-02}$ $d_7 = -2.3963370 \times 10^{-03}$ $d_8 = -8.3823321 \times 10^{-05}$
0.000–42.919	7	$d_0 = 0.000000 \times 10^{+00}$ $d_1 = 1.978425 \times 10^{+01}$ $d_2 = -2.001204 \times 10^{-01}$ $d_3 = 1.036969 \times 10^{-02}$ $d_4 = -2.549687 \times 10^{-04}$ $d_5 = 3.585153 \times 10^{-06}$ $d_6 = -5.344285 \times 10^{-08}$ $d_7 = 5.099890 \times 10^{-10}$

Range (mV)	Order	Coefficients
42.919–69.553	5	$d_0 = -3.11358187 \times 10^{+03}$
		$d_1 = 3.00543684 \times 10^{+02}$
		$d_2 = -9.94773230 \times 10^{+00}$
		$d_3 = 1.70276630 \times 10^{-01}$
		$d_4 = -1.43033468 \times 10^{-03}$
		$d_5 = 4.73886084 \times 10^{-06}$

For all thermocouple types, the tabular data and the polynomial coefficients can be downloaded in machine-readable form from the NIST website.[3]

Reference Junction Compensation

As supplied by the NIST, the relationships for millivolts as a function of temperature are for a reference junction temperature of $0\,°C$. One could immerse the reference junction terminals in a mixture of ice and water to maintain them at $0\,°C$. However, this is not very convenient in an industrial environment.

Figure 2.15 illustrates three thermocouples, all of the same type. Thermocouple 1 is the installed thermocouple, so V is the measured thermocouple voltage for a reference junction at known temperature T_R. But to use the NIST thermocouple tables, we need the value of V_T for a reference junction at $0\,°C$. The computations proceed as follows:

1. Using the NIST tables, determine the millivolts V_R that correspond to the reference junction temperature T_R. This is the electric potential from hypothetical thermocouple 2 in Figure 2.15.
2. Compute the millivolts V_T relative to $0\,°C$ by adding V_R and V. This is the electric potential from hypothetical thermocouple 3 in the figure.

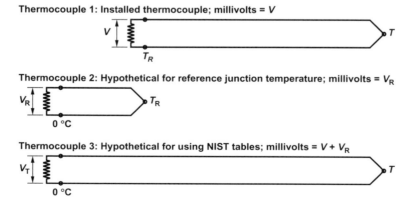

Figure 2.15. Compensation for reference junction temperature.

3. Using the NIST tables, determine the measuring junction temperature T from the millivolts V_T.

Here's a numerical example, suppose the thermocouple voltage V is 8.220 mV for a reference junction temperature T_R of 26.5 °C. The temperature T at the measuring junction is computed as follows (using linear interpolation between the values in Table 2.1):

1. Compute voltage V_R corresponding to T_R. From the data for a type J thermocouple in Table 2.1, a thermocouple with a measuring junction at 26.5 °C and reference junction at 0 °C would generate 1.355 mV (1.329 mV for 26 °C; 1.381 mV for 27 °C).
2. Compute voltage V_T for a thermocouple with a measuring junction at T and a reference junction of 0 °C. This is $V + V_R = 8.220\,\text{mV} + 1.355\,\text{mV} = 9.375\,\text{mV}$.
3. Compute temperature T at measuring junction. From the data for a type J thermocouple in Table 2.1, a thermocouple generating 9.375 mV with a reference junction at 0 °C would have a measuring junction temperature of 174.6 °C (9.337 mV for 174 °C; 9.392 mV for 175 °C).

Thermocouple Types

There are eight thermocouple types, each designated by a letter, intended for industrial applications. Plants using thermocouples generally standardize on a single thermocouple type, so one would use that type unless it is unsatisfactory for a specific reason (for example, the temperature of the application is above the limit for that type of thermocouple).

The selection of the thermocouple type for applications within a facility is based on the following considerations:

Temperature range of the applications. The approximate ranges are as follows:

Type	Positive	Negative	Range, °C	Range, °F
B	Pt-30% Rh	Pt-6% Rh	800–1700	1500–3100
E	Chromel®	Constantan	−200–900	−300–1600
J	Iron	Constantan	0–760	32–1400
K	Chromel®	Alumel®	0–1250	32–2300
N	Nicrosil	Nisil	0–1250	32–2300
R	Pt-13% Rh	Platinum	0–1450	32–2700
S	Pt-10% Rh	Platinum	0–1450	32–2700
T	Copper	Constantan	−200–350	−300–700

The various suppliers of industrial thermocouple probes suggest slightly different ranges, which is why the Celcius to Fahrenheit conversions in the table are not exact.

Chemical resistance to contaminants in the atmosphere. This is difficult to summarize. Types K, S, R, and B should not be used in reducing atmospheres. Consult thermocouple manufacturers for more specific recommendations.

Cost. Type J is the least expensive.

Thermocouple Wires

The following issues pertain to the thermocouple wires.

Grades. The three are precision, standard, and lead wire.

Weight. The larger wire sizes give longer life, especially at higher temperatures. The down size of larger wires is that they conduct more heat away from the measuring junction. Thermowells with a 1/4-in. bore will accommodate #20 gauge wire; larger gauge wires require a thermowell with a 3/8-in. bore.

Polarity. Polarity must be observed when wiring thermocouples. In naming the thermocouples, the positive wire is stated first; the negative wire is stated second. If the temperature of the measuring junction is above the temperature of the reference junction, the input voltage will be positive. If the temperature of the measuring junction is below the temperature of the reference junction, the input voltage will be negative.

Color coding. As used in the United States, the color codes are as follows:

Type	Positive	Color (+)	Negative	Color (−)
B	Pt-30% Rh	Gray	Pt-6% Rh	Red
E	Chromel®	Purple	Constantan	Red
J	Iron	White	Constantan	Red
K	Chromel®	Yellow	Alumel®	Red
N	Nicrosil	Orange	Nisil	Red
R	Pt-13% Rh	Green	Platinum	Red
S	Pt-10% Rh	Black	Platinum	Red
T	Copper	Blue	Constantan	Red

Note that the negative wire is red for all thermocouple types. But don't assume this is universal. In Germany, the positive wire is red; in France, the positive wire is yellow; in Britain, the negative wire is white.

Junction Types

For industrial use, the thermocouple must be packaged in such a manner that it can withstand service conditions. A sheath (or tube) with the thermocouple wires inside provides the required structural rigidity. The diameter of the sheath must be consistent with the gauge of the thermocouple wires and the bore diameter of the thermowell. A spring-loaded fitting is usually included so that the tip of the sheath makes contact with the end of the bore within the thermowell. A head assembly permits the field wiring to be attached to the thermocouple.

There are three options for fabricating the junction:

Grounded junction. The thermocouple wires are basically welded to the sheath at the measuring junction. There is no electrical isolation from the sheath. Because the thermocouple is grounded at the sheath, the circuit must not be grounded anywhere else.

Ungrounded (isolated) junction. The measuring junction does not make electrical contact with the sheath. The lack of a metal-to-metal contact with the sheath results in a slower response time but an improved tolerance to vibrations.

Exposed junction. The junction protrudes beyond the sheath and is fully exposed to the substance being measured. This design is intended for installation without a thermowell, usually because the fastest possible response is desired.

Field Wiring

Low-level signals in an industrial environment are potentially a disaster, as was learned in some early computer installations in the power industry. We know how to properly install thermocouple wiring in a plant. However, there are so many opportunities to make a mistake that we seem to always take a few. Good wiring practices include the following:

- *Use short wiring runs for the low-level signals.* The shortest distances are achieved by installing a transmitter as close as possible to the physical location of the thermocouple. With a transmitter, a current loop or digital transmission can be used for the long runs. The down side is the cost of the temperature transmitter.
- *Avoid power wiring and electrical machinery.* Never run thermocouple signals along with AC wiring of any kind.
- *Use twisted, shielded pairs for the thermocouple extension wires.* The shield must be grounded, but at only one point. The thermocouple extension wires must be of the same materials as the thermocouple itself, but usually of a smaller diameter so as to reduce costs. Most suppliers of thermocouples can also supply the appropriate extension wires.

• *Ground the thermocouple circuit at one and only one point.* Differential input hardware permits the thermocouple to be grounded at the tip. A second ground anywhere in the circuit will result in a *ground loop* (current flowing through the thermocouple wires due to the difference in ground potentials).

Reference Junction   In industrial applications, thermocouples always require reference junction compensation. One approach is to maintain the reference junctions at a known temperature, although not at the ice point. Instead, the isothermal terminal blocks are installed in a cabinet that is heated to a temperature slightly above ambient. Although once common, this is now rare.

The temperature at the reference junction is often sensed using either an RTD, a thermistor, or an integrated circuit (IC) temperature sensor (explained in a subsequent section). But if we have to measure the reference junction with either an RTD or a thermistor, why not just use the RTD or thermistor to measure the process temperature? In some cases, this is perfectly reasonable. But there are two situations in which the thermocouple offers advantages:

• Thermocouples can measure process temperatures higher than either an RTD or a thermistor. In a power plant, hundreds of thermocouples are installed. They have no alternative—furnace temperatures are above the upper limits for RTDs.
• In multipoint thermocouple input systems, using a single isothermal terminal block with multiple terminations means that the reference junction temperature for all thermocouple inputs on that block will be the same. Therefore, only one RTD or thermistor is required, regardless of the number of thermocouples.

Noise   To avoid noise problems, good wiring practices are essential. Most mistakes in the wiring will introduce noise into the system. Although it is tempting to attempt to remove the noise with a filter, this invariably leads to an excessive degree of smoothing. In the end, you will have to locate the wiring mistake and correct it.

When direct thermocouple inputs are used, a component of noise at 50 or 60 Hz will be present. This has to be removed by the input hardware. The problem with digital filters is aliasing. When a signal containing high-frequency noise is sampled at a low frequency, the noise will be present in the digital signal, but at a frequency lower than the sampling frequency.

Special input hardware has been developed specifically for processing the inputs from thermocouples. This equipment uses a *flying capacitor technology* to create what is usually referred to as an integrating A/D converter. After being fully discharged or "shorted," a capacitor is connected to the thermocouple inputs for exactly one cycle AC. The charge on the capacitor reflects the average thermocouple input voltage over this one cycle AC. The capacitor basically integrates the thermocouple input voltage for one cycle AC, which provides excellent rejection of noise at the AC frequency. The capacitor is then switched to the inputs to the A/D converter and the capacitor voltage is read. If one knows the capacitor size and other design parameters of the input system, it is possible to deduce the thermocouple input voltage from the capacitor voltage.

Economics of Thermocouples

An advantage often stated for thermocouples is that they are the most economical way to measure temperature. This statement needs an explanation.

When a temperature transmitter is used, the largest part of the total cost of the measurement system is the temperature transmitter. Thermocouple probes are less expensive than RTD probes, but the probe usually costs about a tenth of the cost of the temperature transmitter. The resulting difference in cost is hardly noticeable.

Some large production units, such as those in oil refineries, require hundreds of temperature measurements, many of which are points such as bearing temperatures. A high percentage of these require neither high accuracy nor high speed. Installing a temperature transmitter on each of these inputs seems like overkill. Before the appearance of digital controls, a system known as a *digital temperature indicator* (DTI) was developed for such applications. The original DTIs consisted of a manual selector switch that the process operator could use to connect any one of a large number of thermocouples to what was basically a digital voltmeter calibrated in temperature units. Today's counterpart is an automatic scanner consisting of a multiplexer that can connect any one of a large number of inputs to an integrating A/D converter. For large point counts, the resulting per-point cost is about one tenth the cost of a temperature transmitter. In large point count applications, the savings are noticeable.

Complete Assembly Specifications

The specifications for the complete assembly include the thermocouple specifications and the thermowell specifications. These usually include at least the following:

Assembly type. Spring-loaded assemblies are probably the most common. Some applications impose other requirements, such as sanitary.

Required measurement range. The type of thermocouple imposes the maximum limits on the measurement range. However, the materials used in some probes impose narrower limits.

Element. The thermocouple can be grounded or ungrounded. For a nominal extra cost (relative to the total cost of the assembly), two thermocouples can be provided in a single probe. For critical measurements, both can be connected to the controls; for others, the spare thermocouple is not connected but otherwise is already installed in case the first one fails.

Head. The head is optional, but usually specified. Various materials of construction are offered. Explosion-proof versions are also available.

Extension. Although not required, this is also usually specified. The configuration is usually nipple–union–nipple with a specified total length. Options are also provided for the material of construction.

Thermowell. Options include immersion length, lagging extension length (if required), style (straight, tapered, stepped), process connection (pipe threads, welded, flanged), and material of construction.

Summary of Advantages

The usual deliberation is thermocouple vs. RTD. Although RTDs are discussed in detail in the next section, we will proceed with stating the advantages of thermocouples:

Applicable to high temperature measurements. Thermocouples can go as high as 1700 °C (3100 °F); RTDs top out at about 850 °C (1560 °F) and thermistors at about 200 °C (400 °F). Because of the higher temperature limit, a lot of thermocouples are currently installed in power plants, and this practice will continue.

Most rugged of contact temperature sensors. Thermocouples are less likely to be affected by vibration and shock.

Most economical for large point counts. The input voltages from thermocouples can be multiplexed, permitting one analog input conversion unit to the applied to many input points. Although such analog input systems are relatively complex, the technology is proven and is widely available.

Fastest response. Unfortunately, this applies to the element itself. When the element is inserted into a thermowell (as most are), the response time is largely determined by the thermowell.

Summary of Disadvantages

Again the contrast is mainly between thermocouples and RTDs:

Low-level signals in an industrial environment. Starting in the 1970s, temperature transmitters were recommended as the best way to avoid long runs of low-level signals in the plant. However, the use of direct thermocouple inputs to controllers, data-acquisition systems, and so on continues to be common practice. The applicable wiring practices are well known. But we do make mistakes, and finding a mistake in thermocouple wiring is rarely easy.

Reference junction compensation is required. The typical user need not be concerned with this. Either the transmitter manufacturer or the data-acquisition equipment manufacturer takes care of it.

Less accurate. Using industrial thermocouple practices, the accuracy of a thermocouple seems to be about 1% of reading. More exact numbers are available for each thermocouple type, and the accuracy is more properly stated as a percent of the difference between the measuring junction temperature and the reference junction temperature. In the power industry, three thermocouples are often inserted into the same location. At 1200 °F, the three temperatures will have a spread in the 10° to 20 °F range.

Subject to drift. Junctions deteriorate with time, especially at high temperatures. The result is drift. The noble metal types drift less than the less-expensive and more commonly used alternatives.

2.6. RESISTANCE TEMPERATURE DETECTORS

The resistance of a metal wire is given by the following formula:

$$R = \frac{rL}{A}$$

where R = resistance of wire (Ω); r = resistivity of the metal ($\Omega \cdot \text{cm}$); L = length of the wire (cm); A = cross-sectional area of the wire (cm^2).

When fabricating resistors from expensive metals such as platinum, there is an incentive to obtain the required resistance using as little metal as possible. This means wires with as small a diameter as possible. But in industrial environments, small-diameter wires can be subjected to vibrations and shock, which can deform or even break the wires. The final result is a compromise between attaining an acceptably robust product and minimizing the amount of the precious metal required to fabricate the resistor.

Effect of Temperature

The resistance of metals increases with temperature. A linear approximation of this relationship is

$$R = R_0(1 + \alpha t)$$

where t = temperature (°C); R = resistance at temperature t (Ω); R_0 = resistance at 0 °C (Ω); α = temperature coefficient of resistance ($\Omega/\Omega \cdot °\text{C}$), usually computed from the resistance at 0 °C and 100 °C.

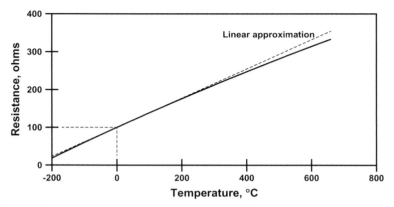

Figure 2.16. Resistance of a 100-Ω platinum RTD as a function of temperature.

However, the relationship between resistance and temperature is not exactly linear. Figure 2.16 shows the relationship for a platinum (RTD-grade) resistor that has a resistance of 100 Ω at 0 °C. The graph also presents the linear approximation using 0.00385 Ω/Ω·°C for the temperature coefficient of resistance. Although many conventional transmitters treated the RTD as linear, linearization is incorporated into the digital transmitters.

The Callendar-Van Dusen equation is one option for expressing the nonlinearity in the resistance–temperature relationship. This equation involves coefficients A and B, which are computed from resistance measurements at 0°, 100°, and 260 °C. A third coefficient C is used only for $t < 0$ °C. The Callendar-Van Dusen equation is as follows:

$$R_t = R_0\left(1 + At + Bt^2 - 100Ct^3 + Ct^4\right)$$

where t = temperature (°C); R_t = resistance at temperature t (Ω); R_0 = resistance at 0 °C (Ω).

The coefficients A, B, and C are defined as follows:

$$A = \alpha + \frac{\alpha\delta}{100}$$

$$B = \frac{-\alpha\delta}{100^2}$$

$$C = \frac{-\alpha\beta}{100^4}$$

where

R_{100} = resistance at 100°C (Ω).

t_h = high temperature $(t_h > 100°C)(°C)$.

R_{t_h} = Resistance at temperature $t_h (\Omega)$.

t_l = low temperature $(t_l < 0°C)(°C)$.

R_{t_l} = Resistance at temperature $t_l (\Omega)$.

$$\alpha = \frac{R_{100} - R_0}{100 R_0}.$$

$$\delta = \frac{100^2 R_0 (1 + t_h \alpha) - R_{t_h}}{R_0 \alpha t_h (t_h - 100)}.$$

$$\beta = \frac{100^4 R_0 \alpha t_l - [100^2 (R_{t_l} - R_0) + \delta t_l (t_l - 100)]}{t_l^3 (t_l - 100)}, \text{ if } t < 0°C; \beta = 0, \text{ if } t > 0°C.$$

Using only the coefficient α, the temperature computed for t_h would be $(R_{t_h} - R_0)/(R_0 \, \alpha)$. The coefficient δ is computed from the difference between temperature t_h and the computed temperature (the error in the temperature computed from R_{t_h}). Similarly, coefficient β is computed from the difference between temperature t_l and the temperature computed from R_{t_l} using only coefficients α and δ.

For an industrial RTD, the values for these coefficients are as follows:

$$\alpha = 0.00385 \, \Omega/\Omega \cdot °C$$

$$\delta = 1.4999 \, °C$$

$$\beta = 0.10863 \, °C, \text{ if } t < 0°C; \beta = 0, \text{ if } t > 0°C$$

$$A = 3.908 \times 10^{-3} \, °C^{-1}$$

$$B = -5.775 \times 10^{-7} \, °C^{-2}$$

$$C = -4.183 \times 10^{-12} \, °C^{-4}, \text{ if } t < 0°C; C = 0, \text{ if } t > 0°C$$

For $t > 0°C$, the Callendar-Van Dusen equation reduces to a quadratic polynomial.

Temperature Coefficient of Resistance

Typical values of α are as follows:

High-purity platinum: $0.00392 \, \Omega/\Omega \cdot °C$.
Platinum used in RTDs: $0.00385 \, \Omega/\Omega \cdot °C$.
Copper: $0.00427 \, \Omega/\Omega \cdot °C$.
Nickel: $0.00627 \, \Omega/\Omega \cdot °C$.
Nickel-Iron: $0.00518 \, \Omega/\Omega \cdot °C$.

Nickel has a temperature coefficient of resistance almost double that of platinum. Therefore, a nickel RTD would have a higher sensitivity to temperature than would a platinum RTD. Higher sensitivity is usually desirable in a sensor. But as Figure 2.17 shows, the nickel RTD exhibits a greater degree of nonlinearity. Furthermore, platinum is usable up to 850 °C, but nickel is usable up to only 260 °C.

Current Source

Resistance cannot be sensed directly. Instead, it is inferred from a voltage.

The simplest approach is to pass a known current through the resistor and measure the voltage drop across the resistor. The circuit shown in Figure 2.18 is a simple approach of this type. A current source is connected in series with the resistor. If the current source supplies 1 ma, the voltage drop across a 100-Ω resistor would be 0.1 V, or 100 mV.

When inferring temperature from the resistance of a metal, there is a down side to passing current through the resistor. The current flow generates heat

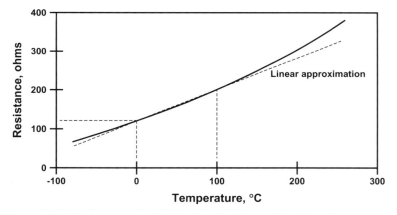

Figure 2.17. Resistance of a 120-Ω nickel RTD as a function of temperature.

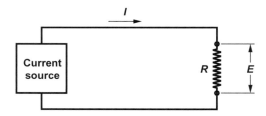

Figure 2.18. Current source in series with a resistor.

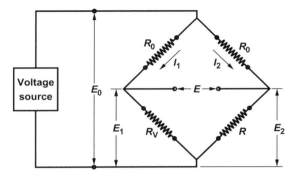

Figure 2.19. Wheatstone bridge.

within the resistor, which increases the temperature of the resistor. Current I flowing through a resistance R generates heat in the amount of I^2R. This is generally referred to as *self-heating*. Because the heat generation increases with the square of the current flow, the need to flow as little current as possible should be obvious.

Bridge Network

Another approach to sensing resistance is to use a Wheatstone bridge as shown in Figure 2.19. A known voltage is imposed on a network consisting of four resistors. Three (R_0 and R_V) are precision resistors; the fourth is the one whose resistance is to be determined. The sensed voltage E depends on the difference between the resistances R and R_V. There are two ways to work with such a bridge:

Balancing mode. Resistance R_V must be a variable resistor. If R_V is adjusted until $E = 0$ (the bridge is said to be *balanced*), then $R = R_V$. If you have worked with a Wheatstone bridge in the laboratory, you probably used this approach.

Nonbalancing mode. The voltage E depends on the difference between R and R_V. For $R_0 \gg R$ and $R_0 \gg R_V$, the current flowing in each arm of the bridge is

$$I = I_1 = I_2 = \frac{E_0}{R_0}$$

The relationships between E and R are

$$E = RI - R_V I = \frac{E_0}{R_0}(R - R_V)$$

$$R = R_V + \frac{E}{I} = R_V + \frac{R_0}{E_0} E$$

Industrial temperature transmitters use the nonbalancing mode approach. In conventional (analog) temperature transmitters, the value of R_V is set equal to the value of R that corresponds to the temperature at the lower-range value of the measurement range. This ensures that E will always be positive.

Equations can be derived without assuming that $R_0 \gg R$. These are more complex, but easily evaluated by a microprocessor. Referring to Figure 2.19, the currents I_1 and I_2 are given by the following expressions:

$$I_1 = \frac{E_0}{R_0 + R_V}$$

$$I_2 = \frac{E_0}{R_0 + R}$$

The voltages E_1 and E_2 are given by the following expressions:

$$E_1 = I_1 R_V = \frac{E_0 R_V}{R_0 + R_V}$$

$$E_2 = I_2 R = \frac{E_0 R}{R_0 + R}$$

The sensed voltage E and the unknown resistance R are related as follows:

$$E = E_2 - E_1 = \frac{E_0 R}{R_0 + R} - \frac{E_0 R_V}{R_0 + R_V} = \frac{R_0 (R - R_V)}{(R_0 + R)(R_0 + R_V)} E_0$$

$$R = \frac{R_0 [E_0 R_V + E(R_0 + R_V)]}{E_0 R_0 - E(R_0 + R_V)}$$

For $R_0 \gg R$ and $R_0 \gg R_V$, these two equations simplify to the equations presented earlier.

The 100-Ω Platinum RTD

Virtually all industrial RTD installations use the 100-Ω platinum RTD that conforms to DIN 43760/DIN IEC 751.[4] Deutsches Institute fur Normung (DIN) is the German industrial standards organization. The 100-Ω platinum RTD standard dates from the 1980s. Another standard is ASTM E1137 *Standards Specification for Industrial Platinum Resistance Thermometers;*[5] however, the practice in the process industries is based almost entirely on the DIN standard.

The choice of platinum as the metal for industrial RTDs is based on several factors:

- Change of resistance with temperature is sufficiently large to be readily measured.
- Departure from linearity is modest.
- Temperature range covers most applications (a major exception is combustion processes, such as in power plants).
- Resistivity of platinum yields resistors of reasonable lengths and acceptable wire sizes.
- Platinum is chemically inert.

The temperature coefficient of resistance of pure platinum is $0.00392\,\Omega/\Omega \cdot °C$. The platinum used for fabricate RTDs is "doped" to give a temperature coefficient of resistance of $0.00385\,\Omega/\Omega \cdot °C$. Industrial RTDs are available in two performance classes (defined in the DIN standard):

Class A. $100\,\Omega \pm 0.06\,\Omega$ or $\pm 0.15\,°C$ at $0\,°C$; usable from $-200°$ to $650\,°C$.
Class B. $100\,\Omega \pm 0.12\,\Omega$ or $\pm 0.30\,°C$ at $0\,°C$; usable from $-200°$ to $850\,°C$.

The majority of industrial applications use Class B RTDs. The resistance-temperature characteristics are the same for both classes.

RTD Temperature Probe

As illustrated in Figure 2.20, an RTD temperature probe is typically constructed from the following components:

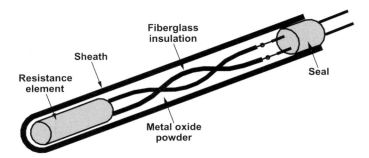

Figure 2.20. Construction of an RTD probe.

RTD element. The wound platinum wire of the RTD is supported by a ceramic housing.

Internal leads. The RTD element is connected to the external leads by wires with fiberglass or other suitable insulation.

Sheath. The sheath provides support and protection. For industrial applications, most sheaths are metal and fabricated for insertion into a thermowell.

Packing. The packing provides both electrical insulation and a conduit for heat conduction. Aluminum oxide and magnesium oxide are commonly used for packing.

Seal. Metal oxides are hydroscopic. The seal is crucial for preventing moisture from entering the probe. Seals may be either epoxy or ceramic.

External leads. These are usually plated copper wires insulated with Teflon.® The probe in Figure 2.20 has only two external leads. But as will be discussed shortly, most probes for industrial service have three leads. A few have four leads.

Although the Class B RTD can be used up to 850 °C, the technology used to fabricate the commonly available RTD probes imposes a limit of about 400 °C. Special and more expensive fabrication techniques are required in order for the probe to be usable up to 850 °C. But because 400 °C is adequate for the vast majority of RTD applications, the less expensive technology is widely used.

Lead Wires or Extension Wires

The extension wires for RTDs are copper wires. Commercially available RTD extension wires contain the required number of wires (as will be explained shortly, most industrial installations use three-wire RTDs) and use the same color coding as for the RTD. The main issue is the electrical resistance provided by the lead wires. This suggests larger-diameter wires would be preferred, which has other advantages in an industrial environment.

In industrial installations, the length of the lead wires can be substantial. The read-out device in Figure 2.21 could be either a current source or a bridge

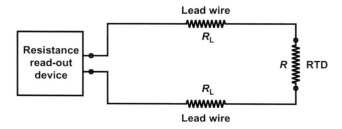

Figure 2.21. Resistance of lead wires to RTD.

network. Either is capable of determining the total resistance of whatever is connected to its terminals. However, the total resistance encompasses.

The RTD element. Resistance is R, which is $100\,\Omega$ more or less, depending on the temperature.

Two lead wires. Resistance of each lead wire is R_L.

Total resistance is $R + 2R_L$. The resistance of the lead wires is a function of

- Length and diameter of the lead wires.
- Temperature of the lead wires.

Although inconvenient, the installation procedures could include a step to determine the resistance of the lead wires. But because this resistance is a function of temperature, the lead wire resistance would not be constant.

Three-Wire RTD

When a bridge network is used, the objective is to develop a circuit so that one lead wire is in each arm of the bridge. The circuit illustrated in Figure 2.22 accomplishes this:

- One lead wire is in series with resistor R_V.
- One lead wire is in series with the RTD (resistance is R).
- One lead wire is in series with the voltage source.

For the case in which $R_0 \gg R_V$ and $R_0 \gg R$, the current I through each arm of the bridge is E_0/R_0. The bridge imbalance voltage is

$$E = I(R + R_L) - I(R_V + R_L) = I(R - R_V)$$

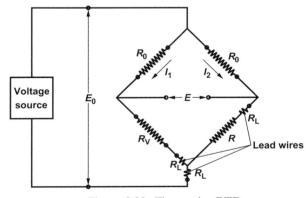

Figure 2.22. Three-wire RTD.

It is important that the lead wire resistances be the same. These lead wires would have the same length and the same temperatures, so there should be very little difference in their the resistances.

Four-Wire RTD

Most industrial applications of four-wire RTDs employ a current source. When a current source is used to pass a known current through the RTD, the voltage at the terminals of the current source would be

$$E_T = I(R + 2R_L)$$

Because the lead wire resistance affects this voltage, the resistance R of the RTD cannot be accurately calculated from this voltage.

In the four-wire RTD configuration used with a current source, two lead wires are connected to each end of the RTD. The wiring configuration is illustrated in Figure 2.23. The current source is connected to the RTD using two of the lead wires. The remaining two are used to sense the voltage across the RTD only. A modern voltage-sensing device has a very large impedance. Because essentially no current will be flowing through these lead wires, the voltage drop across these lead wires will be negligible.

Color Codes

A three-wire RTD will have one white wire and two red wires. The two red wires are connected to one end of the RTD; the white wire is connected to the other end. The two red wires are interchangeable.

Some three-wire RTD temperature probes are dual element—that is, they contain two RTDs. There will be six wires. One RTD is connected via two red wires at one end and one white wire at the other. The second RTD is connected via two black wires at one end and one green wire at the other.

A four-wire RTD for use with current sources will have two white wires and two red wires. The two white wires are connected to one end of the RTD;

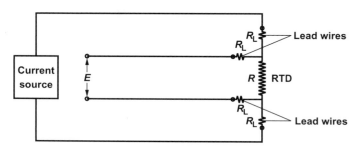

Figure 2.23. Four-wire RTD.

the two red wires are connected to the other end. The two white wires are interchangeable; the two red wires are interchangeable.

Dual element four-wire RTDs are also available. There are two white wires and two red wires for one RTD; there are two black wires and two green wires for the second RTD.

Complete Assembly Specifications

The specifications for the complete assembly include the RTD specifications and the thermowell specifications. These usually entail at least the following:

Assembly type. Spring-loaded assemblies are the most common.

Required measurement range. Although the RTD can sense temperatures up to 850 °C, the materials used in some probes impose narrower limits.

Required accuracy. Class B probes (0.06%) are used in most industrial applications, but for higher accuracy Class A probes (0.12%) are available.

Element. Virtually all industrial applications are three wire. For a nominal extra cost (relative to the total cost of the assembly), two RTDs can be provided in a single probe. For critical measurements, both can be connected to the controls; for others, the spare RTD is not connected but otherwise is already installed in case the first one fails.

Head. The head is optional but usually specified. Various materials of construction are offered. Explosion-proof versions are also available.

Extension. Although not required, this is also usually specified. The configuration is usually nipple–union–nipple with a specified total length. Options are also provided for the material of construction.

Thermowell. Options include immersion length, lagging extension length (if required), style (straight, tapered, stepped), process connection (pipe threads, welded, flanged), and material of construction.

Wire-Wound vs. Thin Film

The traditional approach to manufacturing RTDs is to use long, thin platinum strands that are wound so that the necessary length can be attained in a compact space. This technology is referred to as *wire-wound*. Most industrial RTD applications are of this variety. However, these products are fabricated manually under a microscope.

An alternative technology is to deposit a thin film of platinum on a ceramic base, as illustrated in Figure 2.24. The thin film is then trimmed to give the desired resistance. This technology involves far less manual labor. Furthermore, a resistance of 100 Ω can be achieved with far less platinum. Resistances of 1000 Ω are practical, which provides a greater sensitivity while minimizing the effect of lead wire resistances.

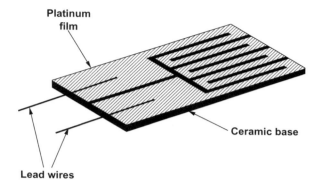

Figure 2.24. Thin film RTD.

Both types of RTDs are widely available. As the wire-wound version is basically handmade, it is likely to eventually go the way of most such products. Although the installed base is largely of the wire-wound variety, the future appears to be with the thin-film technology.

Summary of Advantages

Most chemical plants now prefer RTDs over thermocouples, mainly because of the RTD's superior accuracy and stability. The major advantages of RTDs are

Better accuracy than thermocouples. How much better? At 100 °C, the accuracy of a thermocouple will be about 1 °C (the noble metal thermocouples are somewhat better). Statements on the accuracy of an RTD range from about 0.1° to about 0.5 °C. Both lose accuracy as the temperature increases, but the RTD is always superior to a thermocouple.

Very stable over extended periods. The improved stability is the result of the RTD being constructed from a noble metal. They do not exhibit the drift associated with thermocouples.

Less prone to noise. Being current devices, RTDs are more immune to electrical interference in a plant environment. If someone is smoothing an input from an RTD, questions need to be asked.

Senses temperature, not temperature difference. There is no need for cold junction compensation.

Summary of Disadvantages

The disadvantages are largely relative to thermocouples:

There is no practical way to multiplex RTD inputs. Each RTD requires either a bridge network or a current source to sense the resistance. An individual transmitter can be provided for each RTD, but this is costly. Many data-acquisition systems reduce the costs through RTD cards that accept eight inputs per card. With the cost or the multiple input card being comparable to the cost of a temperature transmitter, the cost reductions are significant in plants with numerous RTDs. But through multiplexers with large point counts, even lower costs are possible when using thermocouples.

Less rugged. RTDs are more susceptible to vibration and shock than are thermocouples.

Lower temperature limits. The upper limit for an RTD is 850 °C (1550 °F). This is too low for combustion processes, such as the boiler or the furnace in a power plant.

Slower response. The thin-film RTDs respond faster than do the wire-wound RTDs. Even so, the thermowell determines the response time in most applications.

More expensive. RTD elements are more expensive than are the thermo-couple elements. The use of thin-film technology has reduced the differential. But when a temperature transmitter is used, the overall cost is not that different.

2.7. THERMISTORS

A sizable number of thermistors (or *therm*al res*istors*) are currently installed in industrial facilities. However, very few are used for process temperature measurements. Almost all are incorporated into other products. For example, most pH probes provide temperature compensation. This temperature is most likely sensed with a thermistor.

The future of thermistors for industrial process temperature measurements hinges on the need for measuring very small temperature changes, such as 0.01 °C. Today, measuring temperature to 0.1 °C is sufficient for almost all industrial applications. However, this is why we use RTDs instead of thermocouples. If we ever need 0.01 °C, we will install thermistors instead of RTDs.

Will we need 0.01 °C in the future? Very few seem to think so, but who knows. In 1970, measuring temperature to 1 °C was acceptable. By the year 2000, measuring to 0.1 °C and controlling to 0.5 °C was prevalent. Why? Because temperature affects all chemical reactions, and the chemists demanded that we do a better job of measuring and controlling the temperatures in industrial reactors. It was really process requirements that drove the change from thermocouples to RTDs in the chemical industry.

Fabrication

Thermistors are sometimes referred to as nonmetallic RTDs. Their fabrication follows typical procedures for ceramics:

1. Mix two or more metal oxides with a suitable binder. The types of oxides and the proportions affect thermistor characteristics.
2. Form to the desired geometry. The two most common thermistor geometries are the bead type and the surface type (disks, wafers, chips, etc.).
3. Dry.
4. Sinter at an elevated temperature. Sintering atmosphere and temperature affect thermistor characteristics.

Bead type thermistors are often sealed in glass to improve their stability. The beads can be as small as a pinhead. Such a small size and the resulting fast response are very appealing in certain applications, such as medical.

Resistance Characteristics

Compared to platinum RTDs, the resistance of thermistors is much greater and is more sensitive to temperature. The nature of the sensitivity is the basis for the following classifications:

Negative temperature coefficient (NTC). The resistance of the thermistor decreases with temperature. The thermistors used for industrial temperature measurement are of this type.

Positive temperature coefficient (PTC). These are sometimes referred to as "switching PTC thermistors" because their resistance increases abruptly at a certain temperature. This makes them ideally suited for initiating actions (such as a shutdown) to avoid equipment damage due to elevated temperatures. One application is to protect the windings in electric motors from thermal damage. However, they have no application to process temperature measurement.

The DIN 43760 standard for the 100-Ω platinum RTD contributed to the industrial acceptance of RTDs for temperature measurement. Unfortunately, no such standard has appeared for thermistors. Thermistors are differentiated by their zero-power resistance, which is the DC resistance at a specified temperature (usually 25 °C) with negligible self-heating.

Resistance-Temperature Characteristic

Figure 2.25 shows a plot of the thermistor resistance as a function of temperature for a specific commercial product (this one was chosen only because the resistance-temperature data could be downloaded over the Internet). This graph of resistance as a function of temperature clearly shows that

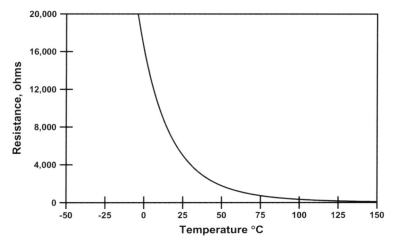

Figure 2.25. Resistance of a commercial thermistor as a function of temperature (R vs. T).

- Thermistor resistance at 25 °C is much greater than 100 Ω. There is an advantage for this: Lead wire resistance is insignificant in comparison. The resistance of the thermistor can be determined using two lead wires and a current source.
- The resistance decreases rapidly with temperature. Thermistors are capable of detecting temperature changes on the order of 0.001 °C. But with such large resistance changes, temperature spans have to be narrow (like 20 °C).
- The relationship is highly nonlinear. This presents great difficulties for analog systems but not for digital systems. The nonlinear problem is often overemphasized. For this reason, we will examine some formulations that very effectively address the nonlinear issues. These will involve logarithms. While this can lead to cardiac arrest for the designers of analog circuits, designers of digital systems barely take notice.

Material Constant, or Beta

Instead of plotting R as a function of T, the presentation is improved by plotting the logarithm of the resistance—that is, $\ln(R)$—as a function of $1/T$, where T is the absolute temperature in degrees K. As illustrated in Figure 2.26, the resulting graph is nearly linear. The material constant of the thermistor, or beta(β) is the slope of the graph. Beta is generally computed from two points. For a temperature transmitter whose measurement range is T_0 to T_1, the following two data points are required:

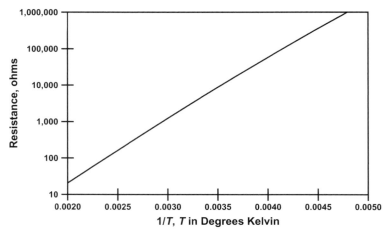

Figure 2.26. Resistance of a commercial thermistor as a function of temperature, plotted as $\ln(R)$ vs. $1/T$.

Thermistor resistance R_0 at temperature T_0 (in degrees K).

Thermistor resistance R_1 at temperature T_1 (in degrees K).

The transmitter computes the temperature T (in degrees K) from the thermistor resistance R using the following equations:

$$\beta = \frac{T_0 T_1 \ln(R_0/R_1)}{T_0 - T_1}$$

$$\frac{1}{T} = \frac{1}{T_0} + \frac{1}{\beta} \ln \frac{R}{R_0}$$

Steinhart-Hart Equation

The use of beta is based on the assumption that $\ln(R)$ varies linearly with $1/T$. However, the graph of $\ln(R)$ vs. $1/T$ is not exactly linear. Consequently, the use of beta introduces error that increases with the temperature span. An alternative is to use the third-order polynomial proposed by Steinhart and Hart:

$$\frac{1}{T} = A + B[\ln(R)] + C[\ln(R)]^2 + D[\ln(R)]^3$$

While somewhat complex, digital systems can advantageously use such equations. The coefficients in this equation can be computed as follows:

From four data points. The values of $A, B, C,$ and D are obtained from the solution of four linear equations.

More than four data points. The values of A, B, C, and D are obtained by applying linear regression. As always, the more data points, the better the results.

Some formulations drop the $C[\ln(R)]^2$ term, which permits the remaining three coefficients to be computed from as few as three data points.

Summary of Advantages

For industrial applications, the advantages of thermistors are as follows:

Can detect small temperature changes. A change of 0.1 °C is about the limit of the capability of an RTD. Thermistors can easily detect a change of 0.01 °C. For most industrial applications, 0.1 °C is sufficient, but who knows what the requirements will be in the future.

Small size. Industrial temperature measurement applications almost always use a thermowell, so extremely small sizes are not required.

Fast response. For probes inserted into thermowells, the thermowell determines the response time. However, occasionally bare probes are used to measure the temperature of a low-pressure gas.

Lead wire resistance is not a problem. Although perhaps a bit inconvenient, three-wire and four-wire RTD installations seem to be manageable.

Summary of Disadvantages

For industrial applications, the disadvantages of thermistors are as follows:

Temperature span must be narrow. Wide spans are needed for start-up but not for normal operations. One could use a thermocouple or RTD for start-up and then use the thermistor for normal operations.

Fragile; avoid shock and vibrations. With proper packaging, this is manageable.

Upper temperature limit lower than for RTDs. There are important industrial processes, such as fermentations, for which the range provided by thermistors is perfectly acceptable.

Exposure to elevated temperatures results in drift. Exposing a thermistor to a temperature above its recommended upper limit usually imparts a permanent error in the readings.

Nonlinear resistance–temperature relationship. With digital systems, this is not really a problem.

2.8. TEMPERATURE TRANSMITTERS

In its simplest form, a temperature transmitter accepts input from either a thermocouple or an RTD and outputs either a 4- to 20-ma current signal or a digital value via a communications interface. Previous discussion has noted the following:

> *Thermocouples.* Installing a temperature transmitter in the field greatly shortens the wiring distances for the low level signals.
>
> *RTDs.* Either a bridge network or a current source is required for each RTD.

Temperature transmitters address both issues.

All modern temperature transmitters are microprocessor based. Therefore, they provide great flexibility. A given transmitter model can be configured to accept any of the following inputs:

> Thermocouples: usually all types.
>
> RTDs: 100-Ω platinum, 1000-Ω platinum (and usually others).
>
> Millivolt: from about -50 to $150\,$mV.
>
> Resistance: up to about $2000\,\Omega$.

On transmitters with dual inputs, both inputs must usually be of the same type.

The list of possible inputs does not generally include thermistors. Temperature transmitters for thermistors are available, but they are specific to thermistors.

Signal Conditioning

All signal conditioning is provided by the temperature transmitter:

> *Thermocouples.* The millivolt input from the thermocouple is compensated for the reference junction temperature and then converted to the process temperature. The reference junction is most often within the temperature transmitter. However, some models provide for the reference junction to be external.
>
> *RTDs.* The external resistance is sensed using a two-, three-, or four-wire configuration. The resistance is then converted to the process temperature.

Most manufacturers can provide NIST-traceable calibrations for their temperature transmitters.

Fail Upscale

Sensors do fail. The obvious consequence is that we do not know the process temperature. But there could be other consequences, especially when the measured temperature is the input to control logic. The likely result is that the controller will drive its control valve either fully open or fully closed. In some

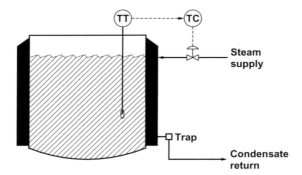

Figure 2.27. Steam-heated vessel.

applications, one of these has serious consequences and must be avoided. For this reason, most temperature transmitters sense the integrity of the sensor element (thermocouple or RTD). On loss of sensor integrity, the transmitter can be configured to fail either upscale (high), downscale (low), or hold last value.

When the interface between the transmitter and the controller is a current loop, the action on failure must be configured within the temperature transmitter. When the interface is a communications interface, the transmitter can report an invalid value to the control system, and the action on failure can be configured within the control system.

In the process in Figure 2.27, the temperature in the vessel is being controlled using steam. On increasing temperature, the controller would close the steam valve. Therefore, the consequences of a sensor failure would be one of the following:

Fail upscale. The controller would fully close the steam valve.

Fail downscale. The controller would fully open the steam valve.

Hold last value. The controller would slowly drive the control valve to either fully closed or fully open, depending on the difference between the controller's set point and the measured temperature at the time the transmitter failed.

For the vessel temperature-control application, fail upscale would usually be preferred (causing the steam valve to close), but there are exceptions. Those familiar with the process must make these decisions; they are properly included in the hazards analysis.

Dual Inputs

Both RTD and thermocouple assemblies can be purchased with dual elements—that is, two identical sensors within the same sheath. The upper-end models of most temperature transmitters will accept dual inputs. Usually they have to be of the same type—that is, two thermocouples of the same type or two RTDs of the same type. The possibilities for such configurations include the following:

- Two measured variables.
- Redundant sensors.
- Temperature difference.
- Average temperature.

The transmitter is usually the most expensive component in the temperature-measurement system. The advantage of using a single temperature transmitter to measure two temperatures should be obvious: It saves money. However, outputting the temperature of each of the sensors is possible only when the transmitter is connected via a communications interface of some type. Temperature transmitters usually have only one current loop output.

Redundant Sensors

Especially when installed in high-temperature applications, thermocouples are prone to failure or burnout. As noted previously, the temperature transmitter can be configured to drive its output either upscale or downscale when this is detected. This is properly viewed as choosing the *least unacceptable* of the alternatives. The other alternative usually leads to a noticeable process upset.

As a thermocouple degrades, its resistance increases. The temperature transmitter senses the resistance of each of the thermocouples. The thresholds are typically as follows:

$500\,\Omega$. Maintenance should replace the thermocouple; however, the temperature can still be sensed.

$2000\,\Omega$. The thermocouple is considered to have failed.

To avoid a process upset on a thermocouple failure, a dual-element thermocouple is installed, and the temperature transmitter is configured for redundant inputs. The idea is to use the better of the two thermocouples as the source of the temperature measurement.

Btu Calculations

Btu calculations are becoming increasingly more common. Let's give an example. An exothermic reaction is proceeding in the vessel illustrated in Figure 2.28. The rate of reaction can be inferred from the heat being removed by the jacket. The steps are

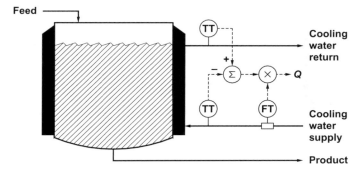

Figure 2.28. Btu calculation.

1. Compute the heat transferred to the jacket from the cooling water flow and the temperature rise.
2. Adjust the heat transfer for the sensible heat of any feed and/or product streams to give the net heat from the reaction.
3. Divide by the heat of reaction to get the reaction rate.

To compute the heat-transfer rate, a temperature difference is required, specifically, the difference between cooling water outlet and inlet temperatures. The configuration in Figure 2.28 provides two temperature transmitters: one for the cooling water inlet temperature and one for the cooling water outlet temperature. These can be subtracted in the software to obtain the temperature difference. Although simple, there are some issues to consider.

Temperature Difference

Numerical analysis courses always emphasize the following: Do not subtract two large numbers to obtain a small number. When one does this, a small error in either of the large numbers results in a large error in the small number. Computing the temperature difference from two temperature measurements does exactly this. Suppose the cooling water enters at 75 °F and leaves at 85 °F, giving a difference of 10 °F. A 1 °F error in either of the large numbers is just over 1%, but a 1 °F error in the temperature difference is 10%.

Most temperature transmitters can be configured to sense the temperature difference from two sensors of the same type. This approach is illustrated in Figure 2.29 and is strongly recommended for applications such as Btu calculations. The advantages are

- Static errors from within the transmitter will cancel. When two separate transmitters are used, any errors in the transmitter zeroes will lead to error in the value computed for the difference.

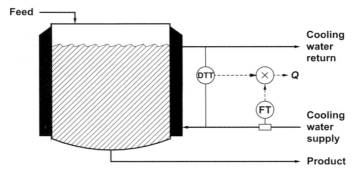

Figure 2.29. Temperature difference as the measured variable.

- Temperature transmitter accuracy is generally stated as percent of upper-range value. The upper-range value will be smaller (and the span narrower) for the temperature difference than for the temperatures, especially for elevated temperatures such as in combustion processes.

2.9. PYROMETERS

In 1894, Stefan developed an equation based on experimental observations; Boltzmann subsequently provided the theoretical basis for that equation. The result is known as the Stefan-Boltzmann Law:

$$E = \varepsilon \sigma T^4$$

where E = total energy emitted by an object (W/cm^2); ε = emissivity; σ = Stefan-Boltzmann constant (5.67×10^{-12} W/cm^2·K^4); T = absolute temperature (K).

When applying pyrometers, there are a couple of points that can be made from this equation. The equation is for energy emitted by the object. A pyrometer needs to receive all energy emitted by the object and only energy emitted by the object. Potential problems include:

- All objects reflect energy, some more than others. A pyrometer is unable to distinguish emitted energy from reflected energy.
- Some gases absorb energy at certain wavelengths (absorption bands). Pyrometers are commonly used in combustion processes because of the high temperatures. The gases in these processes contain water vapor and carbon dioxide, both of which absorb energy at certain wavelengths.

Another observation is that the emitted energy increases rapidly with temperature. An object at 273 °C (546K) is emitting 16 times as much energy as an object at 0 °C (273K).

Planck Radiation Formula

The Stefan-Boltzmann Law relates the total energy emitted to the absolute temperature. The Planck radiation formula describes the distribution of this energy over the various wavelengths, including infrared, visible, and ultraviolet:

$$E_\lambda = \frac{\varepsilon \lambda C_1}{\lambda^5 [\exp(C_2/\lambda T) - 1]}$$

where λ = wavelength (cm); E_λ = energy emitted at wavelength λ (W/cm^2); ε_λ = emissivity at wavelength λ; T = absolute temperature (K); C_1 = constant $(3.74 \times 10^{-12}\,\text{W·cm}^3)$; C_2 = constant (1.44 cm·K).

In this formula, the wavelength is in centimeters; however, wavelength is more commonly expressed in microns, where 1 micron = 1 micrometer (μm) = 10^{-4} cm.

The Planck radiation formula suggests that the energy emitted strongly depends on the wavelength λ:

- λ appears to the fifth power in the denominator. This suggests that more energy is emitted at short wavelengths than at long wavelengths.
- λ appears in exponential term. At short wavelengths, this exponent term is very large. Being in the denominator, this term suggests that the energy emitted at short wavelengths is very small.

These are competing effects. The net result definitely has implications for pyrometers.

Energy Spectrum

Figure 2.30 presents, for various temperatures, the energy emitted at a given wavelength as a function of the wavelength. This graph is computed directly from the Planck radiation formula. The following points are worth noting:

- Each curve in the graph has a peak—that is, a wavelength that corresponds to the maximum emitted energy. The relationship between the peak and the temperature is given by the Wein Displacement Law:

$$\lambda_{max} = \frac{a}{T}$$

where a = 0.290 cm·K. To be most effective, a pyrometer should be responding to the energy emitted at this peak.
- The peak is a function of the temperature. At 0 °C, the peak occurs at a wavelength of approximately 11 μm. At 2000 °C, the peak occurs at a

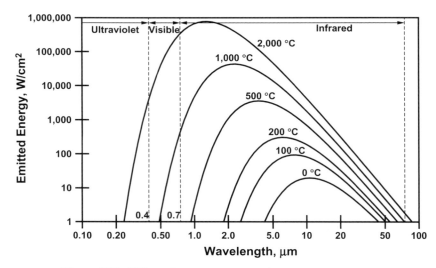

Figure 2.30. Effect of temperature on the energy spectrum.

wavelength of approximately 1.3 µm. Both of these are in the infrared region (0.7 to 80 µm). Pyrometers are usually designed to respond to energy in the infrared region, hence the name *infrared pyrometers*.

- The energy at the peak increases rapidly with temperature. The increase is so large that a logarithmic scale is used in the graph in Figure 2.30. At 0 °C, the peak is slightly less than 20 W/cm²; at 500 °C, the peak is approximately 3,500 W/cm²; at 2000 °C, the peak is almost 800,000 W/cm². Consequently, pyrometer practices for measuring low temperatures tend to be different from pyrometer practices for measuring high temperatures.

Emissivity

Emissivity can be thought of as the efficiency with which an object emits energy. Values for the emissivity are in the range of 0.0 to 1.0. The most efficient emitter is known as a black body and has an emissivity of 1.0. Special designs are required to obtain an object with an emissivity of 1.0. No industrial object is a black body.

Not only is the emissivity of an industrial surface less than 1.0, but the emissivity is also a function of wavelength. A gray body is one whose emissivity is less than 1.0 but is the same at all wavelengths. Some organic materials exhibit behavior close to that of a gray body. However, most objects of industrial interest are non-gray bodies. For these, the emissivity varies with wavelength. A plot of emissivity as a function of wavelength is the *spectral emissivity*.

Tables of the emissivity of various substances are available. The following values are from such a table:

Material	Temperature, °C	Emissivity
Aluminum, unoxidized	25	0.02
	100	0.03
	500	0.06
Iron, oxidized	100	0.74
	500	0.84
	1200	0.89

However, these tables are usually accompanied by disclaimers, such as "The values in these tables should be used only as a guide." Most tables give only a single number, which is usually referred to as the *total emissivity*. It is the mean emissivity over all wavelengths at a given temperature. The values given in the list here are the emissivity at various temperatures, which reflect, among other things, the effect of temperature on the emitted energy spectrum.

Emissivity is always a function of wavelength and temperature. Sometimes other issues arise:

Nature of the surface. For metals, the emissivity is usually distinguished by oxidized vs. unoxidized. Surface preparation, such as polishing, will also affect the emissivity.

Orientation. When the emissivity varies with orientation, the material is said to be anisotropic. Graphite behaves in this manner.

Prior history. Specifically, exposing a metal surface to a high temperature for even a short period of time can change its emissivity.

In summary, emissivity is simply too complex a property to be effectively summarized by a table. Pyrometer manufacturers address this issue by letting the user provide the value for the emissivity. Because different materials have a different emissivity, this is certainly necessary. However, where does the user obtain the value for the emissivity? The difficulty of answering this question continues to plague industrial applications of pyrometers.

Reflected and Transmitted Energy

Pyrometers rely on the fact that the energy emitted by an object is a function of temperature. However, the surrounding objects are also emitting energy to the object of interest. Some of this energy impinges on the surface of interest. There are three possibilities for this energy:

Absorbed by the object. The absorptivity is the fraction of the impinging energy that is absorbed by the object. Like emissivity, the absorptivity is a function of wavelength. The total absorptivity is α; the spectral absorptivity is α_λ.

Reflected by the object. The reflectivity is the fraction of the impinging energy that is reflected by the object. Reflectivity is also a function of wavelength. The total reflectivity is ρ; the spectral reflectivity is ρ_λ.

Transmitted through the object. The transmissivity is the fraction of the impinging energy that is transmitted through the object. Transmissivity is also a function of wavelength. The total transmissivity is τ; the spectral transmissivity is τ_λ.

These must sum to unity:

$$\alpha_\lambda + \rho_\lambda + \tau_\lambda = 1$$

Radiance Temperature

The argument is sometimes made that a given application needs repeatable temperature measurements, but not necessarily accurate temperature measurements. The argument is then made to set the pyrometer's emissivity coefficient to 1.0. The temperature indicated by the pyrometer is the temperature of a black body that would emit the same energy as the object of interest. This temperature is sometimes referred to as the radiance temperature.

Can we maintain constant conditions within the process by simply maintaining a constant radiance temperature? Perhaps in some applications but not when the following arise:

Emissivity. The emissivity is not necessarily constant. If the process temperature were constant, a decrease in the emissivity would give a lower radiance temperature.

Reflectivity. The reflectivity is never 0.0. Consequently, some of the energy received at the pyrometer is emitted by objects other than the object of interest. A change in the temperature of these objects changes the energy they are emitting, which then causes the energy being reflected from these objects to change. Such changes in the reflected energy affect the temperature indicated by the pyrometer, regardless of its emissivity setting.

Opaque Objects

An object is opaque at a given wavelength if no energy with that wavelength is transmitted through the object. For such objects, $\tau_\lambda = 0$. An object is transparent at a given wavelength if all impinging energy with that wavelength is transmitted through the object. For such objects, $\tau_\lambda = 1$. Most transparent objects will have an emissivity close to zero. A pyrometer would sense the temperature of the object that is behind the transparent object. This is great if you are indeed interested in the temperature of that object. But if you are interested in the temperature of the transparent object, you have a problem.

In industrial applications of pyrometers, most objects are opaque at all wavelengths. However, there are exceptions. Consider measuring the temperature of a plastic sheet or film. To the human eye, the sheet may be opaque, it may be transparent, or it may be partially transparent. However, the issue is not how it appears to the human eye, the issue is how it appears to the pyrometer. Materials may be transparent at some wavelengths, but opaque at others.

In this regard, one has to be very careful. An object that is transparent in the visible region may be opaque in the infrared region. The reverse is also possible. One must never assume that the appearance of an object to the human eye will be the same as its appearance to a pyrometer. For example, soda glass is almost completely transparent in the visible region. However, this is not true in the infrared region. For wavelengths beyond 4 μm, soda glass is opaque.

Detectors

Detectors convert radiant energy into an electrical signal. They come in different types:

Thermal detector. An element within the detector is heated by the radiant energy received by the pyrometer. The temperature rise is sensed either by a thermopile (basically several thermocouples in series) or a bolometer (senses the change in resistance of the heated element). Thermal detectors are commonly used when the objective is to measure total energy.

Photon detectors. The photon detector contains a material that releases a charge when exposed to radiation. As compared to a thermal detector, photon detectors are more sensitive and respond more rapidly. However, they have a cutoff—that is, a wavelength beyond which they do not respond. Photon detectors are used in pyrometers that respond to radiation only over a specific range of wavelengths.

Pyroelectric detectors. Ferroelectric materials (lithium tantalite $LiTaO_3$ is one such material) exhibit an electrical polarization that depends on the temperature. Incident radiation affects the temperature, which in turn affects the electrical polarization. This change in polarization is detected through a capacitor that is formed by using a thin slice of the ferroelectric material to separate two electrodes.

Pyrometer Applications

In industry, pyrometers are installed primarily in applications where there is no alternative. The most common such situations are the following:

High Temperatures. The upper limit for thermocouples is about 1700 °C (3100 °F). For higher temperatures, pyrometers are the only option. In fact, the word *pyrometer* is derived from *pyro*, the Greek word for "fire."

Objects are moving. A subcategory of the process industries is the sheet-processing industries, which includes paper machines, carpet lines, and plastic film lines. The heart of these processes is a moving sheet of material. Being a noncontact sensor, the pyrometer is the temperature-measurement device of choice.

Physical contact damages the object. Plastics are extruded at temperatures at which any contact with the surface will leave a defect of some type. Measuring temperatures in such applications can be done only with noncontact technology.

Let's give an example of a temperature measurement for which a pyrometer is the only option. A catalyst is regenerated by burning the carbon from its surface in a fluidized bed. How to you measure the temperature of the catalyst particles in the fluidized bed? The catalyst particles are moving and are at high temperatures, neither of which is a problem for a pyrometer.

Components of a Pyrometer

In its simplest form, a radiation pyrometer contains the three components illustrated in Figure 2.31:

Optics. These are similar to those in telescopes. The radiant energy from the object of interest is focused onto a detector. If appropriate, the object can be very small. Sighting is either through the lens or via a laser pointer that shines a red dot on the target, but occasionally you might encounter a gun sight or a viewfinder.

Detector. This converts the radiant energy into an electrical signal. The choice of the type of detector is left to the manufacturer of the pyrometer.

Filter. Some pyrometers respond to total energy and do not have a filter. If present, the filter passes all energy within a range or band of wavelengths but rejects all energy at wavelengths outside this band.

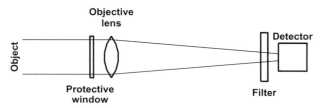

Figure 2.31. Components of a pyrometer.

We will concentrate on the pyrometers for process applications—that is, those that are permanently installed to provide a continuous measurement of temperature. Because the technology is basically the same, most manufacturers also provide portable or handheld versions. These are very convenient for troubleshooting both process and maintenance problems. For example, they are good at locating a hot spot on a wall or vessel.

Low Temperature vs. High Temperature

The application of pyrometers at low temperatures is quite different from their application at high temperatures. Commercial pyrometers are specifically designed for one type of application. A pyrometer designed for high-temperature applications cannot be used in low-temperature applications. A number of factors contribute to this:

- The energy being emitted by the object increases rapidly with temperature. At high temperatures, lots of radiant energy is available to the pyrometer, so it could be designed to respond to only radiation within a narrow band of wavelengths. But at low temperatures, the energy within a narrow band of wavelengths might be less than the pyrometer could detect.
- A pyrometer can sense a temperature of 1000 °C, but electronic equipment does not function at such a temperature (at least not for very long). For low-temperature measurements, the pyrometer can be very close to the object whose temperature is to be measured. For high-temperature measurements, the pyrometer will have to be at some distance.
- At low temperatures, even small amounts of reflected energy or energy absorbed by gases between the pyrometer and the object will introduce large errors. To avoid these, the pyrometer is mounted very close to the surface. Now any variations in the distance will affect the focus. In plastic film applications, the pyrometer must be installed where the sheet passes over a roll, as illustrated in Figure 2.32. Mounting between rolls exposes the pyrometer to sheet flutter.

Broadband Pyrometer

Also called total radiation pyrometers, broadband pyrometers respond to visible radiation plus a large portion of the infrared radiation (typically up to a wavelength of about 20 µm). Based on the curves in Figure 2.30, this covers most of the energy, even though it is not quite total.

For this type of pyrometer, the total emissivity from the usual emissitivity tables apply. But as previously discussed, properly specifying the value for the pyrometer's emissivity coefficient is rarely easy.

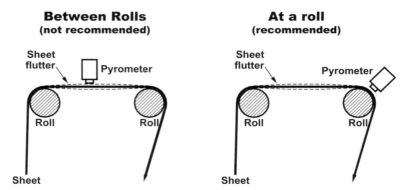

Figure 2.32. Mounting pyrometers for sheet temperature measurement.

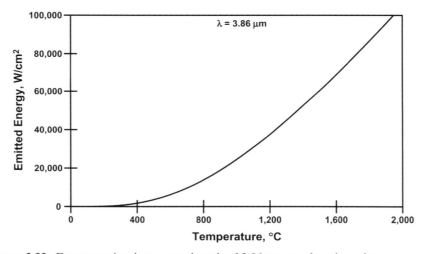

Figure 2.33. Energy emitted at a wavelength of 3.86 μm as a function of temperature.

The Stefan-Boltzmann Law states that the emitted energy varies with the fourth power of the absolute temperature. In practice, better accuracy can be obtained by using a power of N, where N is another coefficient that is specified by the user. The equation for total emitted energy becomes:

$$E = \varepsilon\sigma T^N$$

Narrow-Band Pyrometer

A narrow-band pyrometer responds to the radiation received within a narrow band of wavelengths. Figure 2.33 presents E_λ as a function of temperature for a wavelength of 3.86 μm. This pyrometer would be usable from about 400 °C up.

In combustion applications, one issue that must be addressed is the absorption of radiation by the products of combustion. Any radiation absorbed by the combustion gases does not reach the pyrometer, resulting in a low temperature reading. The major combustion products of concern are water vapor and carbon dioxide. For such applications, the band over which the pyrometer senses the energy must not encompass any of the absorption bands for these compounds.

Ratio Pyrometer

Let the emissivity at wavelength λ_1 be ε_{λ_1} and the emissivity at wavelength λ_2 be ε_{λ_2}. Consider the following:

Black body. $\varepsilon_{\lambda_1} = \varepsilon_{\lambda_2} = 1.0$. Also their ratio $\varepsilon_{\lambda_1}/\varepsilon_{\lambda_2} = 1.0$.
Gray body. Both emissives are less than 1.0. But $\varepsilon_{\lambda_1} = \varepsilon_{\lambda_2}$ and consequently the ratio $\varepsilon_{\lambda_1}/\varepsilon_{\lambda_2} = 1.0$.

Let E_{λ_1} be the energy at wavelength λ_1. Let E_{λ_2} be the energy at λ_2. For specific values of λ_1 and λ_2, the ratio $E_{\lambda_1}/E_{\lambda_2}$ can be computed from the Planck radiation formula and the assumption that the emissivities ε_{λ_1} and ε_{λ_2} are equal. As Figure 2.34 demonstrates, this ratio is a function of temperature. The ratio or two-color pyrometer infers the temperature from this ratio.

Increasing the difference between λ_1 and λ_2 increases the sensitivity of the ratio to the temperature. However, the ratio pyrometer is based on the assumption that $\varepsilon_{\lambda_1} = \varepsilon_{\lambda_2}$, which is more likely to be true when the difference between

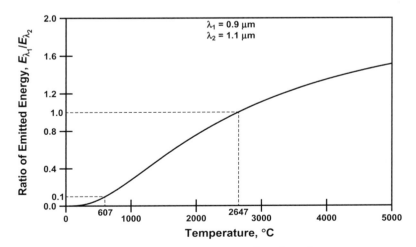

Figure 2.34. Ratio of emitted energy as a function of temperature.

λ_1 and λ_2 is small. At temperatures below about 500 °C, the ratio becomes too small for the pyrometer to sense the temperature.

Optical Pyrometer

The manually operated versions of these project a tungsten lamp filament into the field of view. The operator sights the pyrometer on the object of interest and then adjusts the power to the tungsten filament until it is no longer visible (hence the term *disappearing filament* pyrometers). The tungsten filament is now emitting the same energy as the object of interest. The temperature of the target is inferred from the power being applied to the filament.

By operating in the visible range, optical pyrometers are usable only for measuring elevated temperatures. Measurement ranges typically start at 700 °C (1300 °F). Most use a wavelength of 0.65 μm, which is in the visible red region.

With today's electronics capabilities, it is possible to develop an automated version of the optical pyrometer. But for industrial applications, such a product offers no advantages over a narrow-band pyrometer operating at a wavelength in the infrared region.

Accessories

Although the pyrometer can measure very high temperatures, it cannot survive at these temperatures. The simplest approach is to install the pyrometer at a location sufficiently far from the object that the ambient conditions are within the pyrometer's operating range. But this creates opportunities for interference. Energy can be reflected from other sources. Gases such as carbon dioxide and water vapor can absorb energy. One or more of the following may be required:

Sight tube. Sighting the pyrometer through a tube reduces the radiation from other sources that enters the pyrometer. The sight tube is often purged with air or other suitable gas. This cools the sight tube, reduces the amount of the gases that absorb radiation, and helps to prevent the accumulation of dust on the optics.

Air-purged enclosure. Even without a sight tube, air purging should be considered. Maintaining the pressure within the pyrometer slightly above atmospheric reduces the contaminants and associated corrosion. Directing an airflow across the lens surface helps keep the optics clean.

Water jacket. With this, the pyrometer can be installed when the temperature is up to about 400 °C (750 °F). The down side is that a water jacket is a form of active cooling. It works very well provided the water flow is maintained. In practice, this proves to be a much bigger problem than most expect. If active cooling is ever interrupted, you will be buying another pyrometer.

Calibration

Pyrometers are normally calibrated using black-body simulators whose design depends on the operating temperature range of interest. Most pyrometer manufacturers can provide a calibration certificate that traces their black-body calibration back to NIST standards.

But industrial objects are not black bodies. All pyrometers provide a coefficient that the user can adjust to account for this. The significance of the coefficient depends on the type of pyrometer:

Broadband (total radiation) pyrometer. Coefficient is the total emissivity.

Narrow-band pyrometer. Coefficient is the emissivity at the wavelength used by the pyrometer.

Ratio pyrometer. Coefficient is the *emissivity slope*, or the ratio of the emissivities at the two wavelengths used to compute the ratio.

Setting these coefficients is never a trivial task. What you really need is the emissivity spectrum for the object whose temperature you are measuring (for nonopaque objects, you also need the absorption spectrum). Laboratories can measure both. This obviously requires time and money, but if you make the effort to understand your application, you will probably succeed. Otherwise, you might get lucky, but probably not.

When a secondary temperature measurement is available, it is tempting to perform a single-point calibration by simply adjusting the pyrometer's emissivity coefficient until the temperature indicated by the pyrometer agrees with the other measurement. Laboratories can also perform such calibrations, usually at more than one point and with better precision and documentation. But for some high-temperature processes, the only temperature measurement you have is that from the pyrometer.

2.10. OTHERS

We will briefly discuss a few other temperature-measurement technologies, some that have been around for years and some that are relatively recent.

Industrial Glass Thermometers

Industrial versions of glass thermometers continue to be available. Their low cost is offset by the obvious issues associated with the presence of glass in a production area. Approaches can be taken to enhance their resistance to impact, shock, and vibration. These reduce the probability of a breakage, but with glass, one must always be prepared to deal with the consequences of a break.

Some industrial glass thermometers continue to use mercury as the liquid. Mercury is toxic, and even a small amount of mercury in the wrong place is quite serious. The use of an organic liquid in lieu of mercury avoids these issues.

Capillary Tubes

Sometimes called Bourdon tube thermometers, capillary systems infer the temperature by using a Bourdon tube to sense the pressure in a closed system consisting of a bulb and a capillary tube. There are three varieties of capillary systems:

Gas filled. The temperature–pressure relationship is known from the gas law equations. Air is often used as the filling gas.

Vapor filled. The bulb is partially filled with a liquid, with the vapor from this liquid occupying the remaining volume. The pressure will be the vapor pressure of the liquid at the current temperature. When installing these types, the bulb orientation must be such that only vapor enters the capillary tube.

Liquid filled. The Bourdon tube responds to a change in the liquid volume resulting from a change in temperature. The coefficient of expansion of the liquid determines how much the liquid volume changes with temperature.

A capillary tube thermometer is illustrated in Figure 2.35. The major advantage of the capillary systems is that the capillary tube is very flexible and can be quite long (even 100 ft. is possible). Capillary tube arrangements continue to be used in the following products:

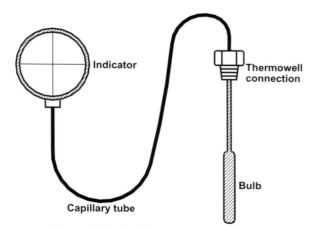

Figure 2.35. Capillary tube thermometer.

Dial thermometers. Most dial thermometers are bimetallic. But whereas the indicator (the dial) must be in close physical proximity to the helical coil, using a capillary system permits the dial to be located some distance from the bulb.

Switches. That the bulb can be some distance from the physical location of the switch is the primary advantage of a capillary system over a bimetallic system. It also requires no external source of power to function.

Pneumatic instrumentation and controls. When this technology was dominant in process controls, capillary systems were the most common method for sensing temperature. Although far less common, pneumatic systems have not entirely disappeared.

IC Temperature Sensors

IC temperature sensors are incorporated into a number of products; users of these products are often unaware of their existence. One example of their use is in reference junction compensation circuits for thermocouples. The IC temperature sensor is limited to the temperature range of integrated circuits (approximately $-55°$ to $125°C$).

The IC temperature sensor generates an output that is directly proportional to the absolute temperature. The following equation for a diode shows that the relationship between voltage and current depends on the absolute temperature:

$$V_F = \frac{KT}{q} \ln \frac{I_F}{I_S}$$

where I_F = forward current (amp); I_S = saturation current (amp); k = Boltzmann constant (1.38×10^{-23} J/K); q = electron charge (1.6×10^{-19} C); T = absolute temperature (K); V_F = forward voltage across diode (V).

Practical implementations often use two diodes (actually transistors that operate as diodes) that are subject to different current flows. The difference in voltage is a function of the absolute temperature. This output voltage is then amplified to give a sensitivity of either 10 mV/°C or 10 mV/°F. Because the IC temperature sensor is an integrated circuit, it is quite easy to incorporate a D/A converter directly into the IC to give a digital output.

Optical Fiber Temperature Sensors

Advancements in measurement technology are often driven by difficult applications. One such application is sensing temperatures within transformer windings. Multiple temperature probes must be inserted at the time the transformer is manufactured; thereafter, the probes cannot be extracted for purposes such as calibration. During transformer operation, the probes will be subjected to strong magnetic fields. The objective is to operate the transformer with as high a power load as possible without exceeding the temperature limits for the transformer windings.

Optical technology can measure a length with great precision. The Fabry-Perot displacement sensor can be applied to the measurement of temperature, pressure, vibration, and others. Temperature measurement is based on the thermal expansion characteristics of a glass element (for example, the length measurement could be the inside diameter of a glass tube).

Other possibilities are specific for the measurement of temperature. In one approach, light is passed through a gallium arsenide semiconductor crystal. The transmission spectrum (the transmission of light as a function of wavelength) depends on temperature. In another approach, light excites a photoluminescent sensor, and the decay is measured. The decay is a function of temperature.

In applications such as transformer windings, a major advantage of fiber optics is its immunity to electromagnetic interference. Optical fibers are initially being used in such applications because there is no alternative. But as the technology matures, fiber optic technology will be commonly applied to a wide range of measurement applications.

Regardless of the variable being sensed, the advantages of optical fiber sensors include the following:

- Immune to electromagnetic interference.
- Can measure practically any physical quantity.
- Large number of principles can be applied.
- Capable of high sensitivities.
- Probes are small and flexible.
- Single readout can have many fiber optic inputs.

Acoustic Pyrometer

Another difficult measurement is flue gas temperature. Flue gas temperature is one of the measurements required to compute the thermal efficiency of a combustion process. A combustion process results in a gas stream at a temperature of 2500 °F, more or less. While within the measurement range of a thermocouple, the following factors lead to measurement errors that are difficult to quantify:

- Frequent recalibration due to thermocouple drift.
- Radiant heat transfer is significant.
- Gases have poor heat-transfer coefficients.
- Probe must be at appropriate location.

The acoustic pyrometer is a noncontact temperature-measurement device that infers the gas temperature from the speed of sound within the flue gas. A transmitter mounted on one side of the flue gas duct outputs a pulse of sound that is detected by a receiver mounted on the other side of the duct. The sonic velocity in the flue gas is computed from the travel time for the pulse of sound and the distance between transmitter and receiver. Knowing the sonic velocity, the flue gas temperature can be computed using the following equation:

$$T = \frac{Mc^2}{\gamma R}$$

where c = sonic velocity within the gas; R = gas law constant; T = absolute temperature; M = molecular weight of the gas; γ = ratio of specific heats (c_P/c_V); c_P = specific heat at constant pressure; c_V = specific heat at constant volume.

The transmitter and receiver need not be exposed to the flue gas itself, making this a noncontact temperature-measurement system. One advantage of this approach is that the acoustic pyrometer measures the average flue gas temperature along the acoustic path. If necessary, additional transmitters and receivers can be installed to provide multiple paths.

One source of error is in measuring the travel time. The travel time is made even shorter by the fact that the sonic velocity increases with temperature. Another source of error is that the molecular weight of a gas depends on the gas composition. For flue gases, the gas composition depends on the type of fuel, the fuel to air ratio, etc.

LITERATURE CITED

1. Murdock, J. W., "Power Test Code Thermometer Wells," *Journal of Engineering for Power, Transactions ASME*, **81**, 1959.

2. American Society of Mechanical Engineers, *ASME/ANSI Power Test Code* 19.3, ASME, New York, 1974 (reaffirmed 1986 and 1998).

3. Downloaded from the NIST website using the following URL: http://srdata.nist.gov/its90/download/type_j.tab

4. Deutsches Institut fur Normung (German Industrial Standards Organization), *Industrial platinum resistance thermometer sensors*, DIN EN 60751, DIN, Berlin, 1996.

5. American Society for Testing Materials, *Standard Specification for Industrial Platinum Resistance Thermometers*, ASTM E1137, ASTM, West Conshohocken, PA, 1997.

CHAPTER 3

Pressure

Applications requiring the measurement of process pressure include the following:

- Many pressure processes perform most efficiently at the highest pressure permitted by the process equipment. Based on a measurement of the process pressure, constraint control logic can maintain the process operating conditions close to the upper pressure limit.
- Some chemical reactors operate at high vacuum (an absolute pressure measurement), whereas others operate either just above atmospheric pressure (no air leaks in) or just below atmospheric pressure (no chemicals leak out).
- Towers and columns must be operated so that the pressure drop across each section (packed or trays) is less than the maximum allowable pressure drop to avoid flooding.

In addition to process pressure measurements, pressure or differential pressure is used as the basis for:

Level. About 75% of the level measurements rely on pressure or differential pressure measurements of the liquid head.

Flow. Orifice and other head-type flow meters require a measurement of the differential pressure. Orifice meter installations in the chemical industry are on the decline, but other industries continue to rely on the orifice meter as their flow meter of choice. In either case, the installed base is quite large.

Physical properties. Density and viscosity measurements are often based on pressure or differential pressure.

3.1. FORCE AND PRESSURE

Over the centuries, the understanding of force was a key component of the advancement of science. Sir Isaac Newton was a key contributor:

Gravitational force. Two objects attract each other with a force that is proportional to their masses and inversely proportional to the square of the distance between them.

Acceleration. Force equals mass times acceleration.

In recognition of Newton's contributions, the unit of force in the SI system is the Newton.

The following industrial measurements involve force:

- Pressure.
- Weight.
- Strain or deformation.
- Torque.

Force is actually a vector quantity. That is, it has a magnitude and a direction. Although not really an issue in pressure measurements, attention must be directed to this aspect in all weight measurements.

Units for Force

In the English system of units, force is measured in pounds. Unfortunately, the pound is used for both mass and force. To distinguish, we will use pounds-mass (lb_m) for mass and pounds-force (lb_f) for force. They are related in this way: $1\,lb_f$ is the force exerted by gravity on $1\,lb_m$. The acceleration due to gravity is $32.2\,ft/sec^2$. Applying Newton's law of acceleration:

$$1\,lb_f = 32.2\,lb_m \cdot ft/sec^2$$

In the SI system of units, force is measured in Newtons. A Newton (N) is the force required to accelerate a mass of $1\,kg$ by $1\,m/sec^2$:

$$1\,N = 1\,kg \cdot m/sec^2$$

The acceleration due to gravity is $9.81\,m/sec^2$. Therefore, the force exerted by gravity on a 1-kg mass is $9.81\,N$.

Gravitational Constant

Equations involving force often include a quantity referred to as the gravitational constant g_c. This is nothing more than a conversion factor! We do not normally include conversion factors for the engineering units in our equations. But for equations involving force, we seem to make an exception. For example, the weight w of an object of mass m is

$$w = mg$$

where g is the acceleration due to gravity (32.2 ft/sec^2 or 9.81 m/sec^2). However, this equation is frequently written as follows:

$$w = \frac{mg}{g_c}$$

The origin for this seems to be associated with the use of the English systems of units, where the pound is used for both mass and force. Usually the best way to deal with this is to avoid pounds, but use either pounds-mass (lb_m) or pounds-force (lb_f) instead. Thus psi is really lb_f/in^2.

Pressure

A common approach for introducing the subject of pressure is the hydraulic cylinder illustrated in Figure 3.1. A weight is placed on a piston of known area. As pressure is force per unit area, the resulting pressure is

$$P = \frac{F}{A}$$

where F = force (lb_f or N); A = area (in^2 or m^2); P = pressure [lb_f/in^2 (psi) or N/m^2 (Pascal)].

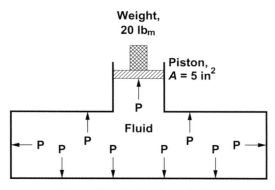

Figure 3.1. Hydraulic cylinder.

If a mass of $1.0\,lb_m$ is placed on a cylinder with an area of $1.0\,in^2$, a pressure of $1.0\,lb_f/1.0\,in^2 = 1.0\,psi$ is exerted throughout the hydraulic fluid. If a mass of $1.0\,kg$ is placed on a cylinder with an area of $1.0\,m^2$, the pressure exerted throughout the hydraulic fluid is

$$\frac{1.0 \times 9.81\,N}{1.0\,m^2} = 9.81\,N/m^2 = 9.81\,Pascals$$

Units for Pressure

The normal units for pressure are as follows:

English system of units. Pounds per square inch (psi) or pounds-force per square inch (lb_f/in^2). This is used almost universally in plants within the United States.

SI system of units. Pascal (Pa), which is equal to $1\,N/m^2$; 1 Pa is a very low pressure (a 1-kg weight placed on a piston of $1.0\,m^2$ gives a pressure of 9.81 Pa). Consequently, most pressure measurement devices use the kilo-Pascal (kPa), which is 1000 Pa. This is widely used in plants in Europe. The bar, which is 100 kPa, is also in common use.

You will encounter a number of other units. Some conversion factors are as follows:

1 psi =	1 kPa =	1 bar =	1 mm Hg =	1 in Hg =	1 in H₂O =	1 atm =
psi	0.14504 psi	14.504 psi	0.019337 psi	0.49115 psi	0.036126 psi	14.696 psi
8948 kPa	1 kPa	100 kPa	0.13332 kPa	3.3864 kPa	0.24908 kPa	101.33 kPa
068948 bar	0.01 bar	1 bar	0.0013332 bar	0.033864 bar	0.0024908 bar	1.0133 bar
.715 mm Hg	7.5006 mm Hg	750.06 mm Hg	1 mm Hg	25.400 mm Hg	1.8683 mm Hg	760.00 mm Hg
0360 in Hg	0.29530 in Hg	29.530 in Hg	0.039370 in Hg	1 in Hg	0.073689 in Hg	29.921 in Hg
7.681 in H₂O	4.0147 in H₂O	401.47 in H₂O	0.53525 in H₂O	13.595 in H₂O	1 in H₂O	406.79 in H₂O
068046 atm	0.0098692 atm	0.98692 atm	0.0013158 atm	0.033421 atm	0.0024582 atm	1 atm

The basis for millimeters of mercury (mm Hg) and inches of mercury (in Hg) is mercury at $0\,^{\circ}C$; the basis for inches of water (in H_2O) is water at $4\,^{\circ}C$.

A few observations:

- The atmosphere (atm) is based on a mean atmospheric pressure of 1 atm.
- For the bar, 1 bar is slightly less than 1 atm. Although commonly used in Europe, expressing pressure in bars is unusual in industrial plants in the United States. However, the central pressure of hurricanes is now routinely stated in millibars instead of inches of mercury. Normal

atmospheric pressure is 1013 mbar; hurricane Camille came ashore at 909 mbar, which is a little over 10% below mean atmospheric pressure.

- The unit torr is often used in lieu of millimeters of mercury (1 mm Hg = 1 torr).

3.2. MEASURES OF PRESSURE

Let's begin by defining the following two pressures:

Absolute zero pressure. Pressure present when all free molecules have been removed from a volume (that is, a perfect vacuum).

Atmospheric pressure. Pressure produced by the effect of gravity on the molecules in the atmosphere. The generally accepted value for atmospheric pressure at sea level is 760 mm Hg = 101.325 kPa = 14.696 psi = 1 atm.

Bases for Expressing Pressure

In industrial applications the measures for pressure are as follows:

Absolute pressure. Pressure relative to absolute zero pressure. When pressure is measured in psi, absolute pressure is usually designated *psia* for "psi absolute."

Gauge pressure. Pressure relative to atmospheric pressure. When pressure is measured in psi, gauge pressure is usually designated *psig* for "psi gauge."

Vacuum. Pressure values that are below atmospheric pressure (negative gauge pressures).

Differential pressure. Difference between two pressure measurements. The two pressure measurements can be absolute pressure or the two pressure measurements can be gauge pressure. The units on differential pressure are psi.

Figure 3.2 illustrates the relationship between these.

Absolute and Gauge Pressure

For processes operating under vacuum, pressure measurements are in absolute pressure. Otherwise, pressure measurements are usually gauge pressure. Gauge pressure and absolute pressure are related by the following equation:

$$P_{gauge} = P_{abs} - P_{atm}$$

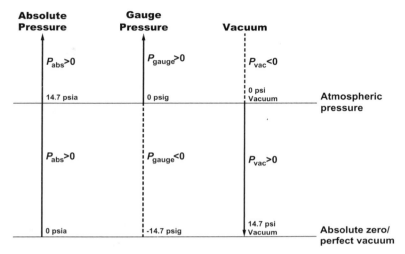

Figure 3.2. Measures of pressure (in psi units).

where P_{gauge} = gauge pressure (psig); P_{abs} = absolute pressure (psia); P_{atm} = atmospheric pressure (14.696 psia).

Occasionally pressure is expressed as *vacuum*. This is simply the negative of gauge pressure. Gauge pressures above atmospheric pressure are positive; gauge pressures below atmospheric pressure are negative. Vacuum is just the opposite. Pressures below atmospheric pressure are positive vacuum; pressures above atmospheric pressure are negative vacuum (these are rarely expressed as "vacuum").

Hydrostatic Head

The water tank illustrated in Figure 3.3 sits on a hill. The tank is open to the atmosphere, so the pressure at the surface of the water is 0 psig. There are three pressure measurements:

At tank discharge. The tank discharge is 15 ft below the surface of the water.
At upper valve. This valve is 40 ft below the tank discharge.
At lower valve. This valve is 40 ft below the upper valve.

A hydrostatic head is a pressure resulting from a column of liquid. One way to express the pressure at these locations is in feet of the liquid—in this case, in feet of water. At the tank discharge, the hydrostatic head is 15 ft of water. At the upper valve, the hydrostatic head is 40 + 15 = 55 ft of water. At the lower valve, the hydrostatic head is 40 + 40 + 15 = 95 ft of water.

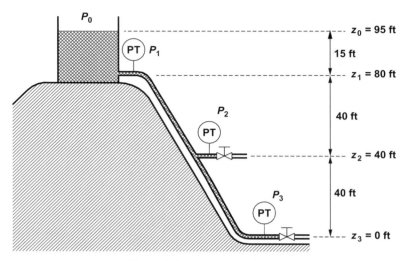

Figure 3.3. Hydrostatic head.

Bernoulli Equation

The Bernoulli equation provides the basis for converting a hydrostatic head to a pressure. When there is no flow, the Bernoulli equation reduces to the following:

$$\frac{P_0}{\rho} + gz_0 = \frac{P}{\rho} + gz$$

$$P - P_0 = \rho g(z_0 - z)$$

$$P = \rho g(z_0 - z) = G\rho_{water}\, g(z_0 - z) \quad \text{for} \quad P_0 = 0$$

where z = elevation (ft); P = pressure at elevation z (psig); P_0 = pressure at elevation z_0 (for the water tank, $P_0 = 0$ psig); g = acceleration due to gravity (32.2 ft/sec^2); G = specific gravity of liquid relative to water (for water, $G = 1.0$); ρ = density of liquid (lb$_m$/ft^3; for water, $\rho = 62.4$ lb$_m$/ft^3) ρ_{water} = density of water (62.4 lb$_m$/ft^3 or 1 g/cc or 1000 kg/m^3).

At the tank discharge, $z_0 - z_1 = 15$ ft. The pressure is computed as follows:

$$
\begin{aligned}
P_1 &= \rho g(z_0 - z_1) \\
&= \frac{62.4\,\text{lb}_m}{\text{ft}^3} \times \frac{32.2\,\text{ft}}{\text{sec}^2} \times (15\,\text{ft}) \times \frac{1\,\text{ft}^2}{144\,\text{in}^2} \times \frac{\text{lb}_f \cdot \text{sec}^2}{32.2\,\text{ft} \cdot \text{lb}_m} \\
&= 6.5\,\text{psig}
\end{aligned}
$$

By similar calculations, $P_2 = 23.8\,psig$ and $P_3 = 41.2\,psig$.

Sometimes this relationship is expressed in terms of the differences ΔP and Δz:

$$\Delta P = \rho g \Delta z = G\rho_{water} g \Delta z$$

where ΔP = difference in pressure, expressed in head of liquid; Δz = Difference in elevation.

Liquid Head

A hydrostatic head can be converted to a pressure. Conversely, any pressure can be expressed as a head of liquid. The liquid head is the height of liquid that would exert a given pressure. The most commonly used liquids are the following:

Water, usually as inches of water but occasionally feet of water.

Mercury, usually as millimeters of mercury (or torr) but occasionally as inches of mercury.

For the water tank, the pressures can be stated as follows:

$$P_1 = 6.5\,psig = 15\,ft \text{ of water}$$
$$P_2 = 23.8\,psig = 55\,ft \text{ of water}$$
$$P_3 = 41.2\,psig = 95\,ft \text{ of water}$$

The density of a liquid is a function of temperature. When expressing liquid head, one should specify the liquid temperature. For mercury, the temperature is usually $0\,°C$. For water, it is usually $4\,°C$ (water is most dense at $4\,°C$), but in the petroleum industry $60\,°F$ is frequently used.

Barometer

A barometer is an absolute pressure measurement device. A glass tube is completely filled with mercury and then inverted into a reservoir of mercury. The space above the mercury in the glass tube is as close to a perfect vacuum as we can achieve. Actually, the pressure in this space is the vapor pressure of mercury. Although not exactly zero (vapor pressure of mercury at $0\,°C$ is about $0.0002\,mm\,Hg$), it is as close to zero as we will get.

The height of the column of mercury depends on the atmospheric pressure. At mean atmospheric pressure, the column of mercury rises $760\,mm$ ($29.921\,in$) above the surface of the mercury in the reservoir.

In the 1600s, pump makers in Florence attempted to build a pump that would raise water 40 ft at its suction. To explain why this effort failed, Evangelista Torricelli, a follower of Galileo, developed the mercury barometer for measuring atmospheric pressure. A unit of measure for pressure is named after Torricelli; as mentioned earlier, 1 torr is equal to 1 mm Hg.

Although industrial issues led to the development of the barometer, it has no industrial applications today. However, the mercury barometer is still used to measure atmospheric pressure in weather stations.

Manometer

A manometer is a differential pressure measurement device. A glass U-tube is partially filled with a liquid. One end is connected to the space whose pressure is to be sensed. The other end is usually open to the atmosphere. The difference in the levels of the liquid in the U-tube reflects the difference in the pressure at the two ends of the U-tube. In the configuration shown in Figure 3.4, the pressure in the space would be slightly above atmospheric pressure. If the pressure in the space is slightly below atmospheric pressure, the connections should be reversed.

When one end is open to the atmosphere, the difference in the levels reflects the gauge pressure—that is, the difference between the pressure in the space and atmospheric pressure. To measure the difference in pressure between two spaces, the ends are connected to the spaces.

A common requirement is to maintain the pressure in a process vessel either slightly above atmospheric pressure (so no air leaks in) or slightly below

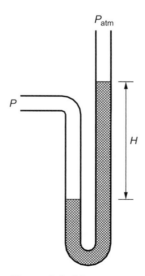

Figure 3.4. Manometer.

atmospheric pressure (so no process gases are released to the atmosphere). In the 1950s, manometers were commonly installed in industrial plants to measure such small pressure differences. To avoid broken glass, the tubes were plastic. For reading small pressures and to avoid the potential of mercury contamination, the liquid was usually water but sometimes an alcohol. However, excessive pressure differentials would "blow" the manometer fluid, possibly into the process. Today, manometers are rarely encountered in industrial applications.

3.3. PRESSURE-SENSING ELEMENTS

The measurement of pressure basically involves two steps:

Translate the pressure into a mechanical displacement. Three alternatives for translating pressure into a mechanical displacement are as follows:
- Bourdon tube.
- Bellows.
- Diaphragm.

Sense the mechanical displacement. In modern pressure-measurement devices, the mechanical displacement is sensed through its influence on an electrical property. This permits a smaller displacement to be accurately sensed, which has led to greater use of diaphragms.

All of these mechanical elements have been around for many years. Advancements in metallurgy have led to some improvements. However, the greatest improvement has been in the application of electronic and then microprocessor-based technologies to sense the displacement and perform the computations required to obtain the measured value for pressure.

Bourdon Tube

In 1832 the French engineer Eugene Bourdon devised what is today known as a Bourdon tube. Although Bourdon tubes now come in various geometries, all are bent or curved hollow tubes with an elliptical cross-section. The tube is sealed at one end. When pressure is applied to the open end, the tube straightens, thus converting pressure into a displacement.

The most common geometries of Bourdon tubes are illustrated in Figure 3.5. The simplest is the C-shaped tube. However, this tube gives the smallest displacement and is generally limited to lower pressures. For pressures up to about 50 psig, the Bourdon tube can be made of metals such as copper; steel is used for higher pressures. The spiral and helical Bourdon tubes can be used up to about 100,000 psig.

Bourdon tubes generally display hysteresis. That is, for a given process pressure, approaching the pressure from below gives one value for the process pressure, but approaching the pressure from above gives a slightly different

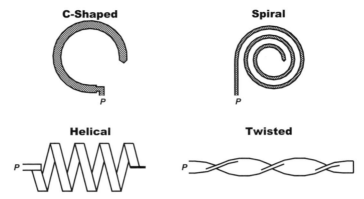

Figure 3.5. Bourdon tube configurations.

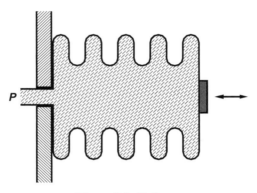

Figure 3.6. Bellows.

value. The typical specification for hysteresis is "not to exceed 0.1% of full scale value." In some services, the Bourdon tube is susceptible to plugging. In liquid service, the Bourdon tube can also trap gases, so some products provide a bleeder port.

Bellows

As illustrated by the bellows in Figure 3.6, the sides expand with increasing pressure and contract with decreasing pressure. Most bellows used in industrial pressure-measurement devices are metallic. Sometimes a spring is used in conjunction with the bellows. The bellows has two advantages:

Sensitive to low pressures. It can sense very small pressures, both slightly above and slightly below atmospheric pressure. An example of a range might be from −5 in H_2O to +5 in H_2O.

Provides power for driving recording and indicating mechanisms. In the days of pneumatic instrumentation, this was a significant advantage. But with electronic instrumentation, this is no longer the case.

The maximum pressure that can be sensed with a bellows is about 1000 psig. To do so, the bellows must be coupled with a heavy spring mechanism.

Differential pressure can be sensed by arranging two bellows in an opposing manner.

Diaphragm

As used in industrial instrumentation, a diaphragm is usually a flexible metal disk. Increasing pressure causes the diaphragm to deform. The diaphragm is the pressure-sensing mechanism of choice in electronic instrumentation because of the following characteristics:

Compact arrangements are possible. The small deformation of the diaphragm is a problem for pneumatic and mechanical systems but is no problem for electronic systems.

Diaphragms can sense gauge pressure, absolute pressure, or differential pressure. As illustrated in Figure 3.7, all use the same basic mechanism.

Diaphragms respond to negative pressure differentials as well as positive pressure differentials. Basically, the only difference is the direction in which the diaphragm deforms.

Diaphragms can be constructed from a variety of materials. Pressure transmitters with thin diaphragms can have a very narrow span, such as −5 in H_2O to +5 in H_2O. Thick diaphragms can sense pressures of 100,000 psig.

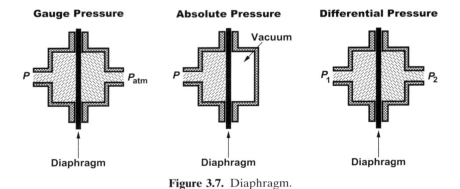

Figure 3.7. Diaphragm.

3.4. INDICATORS AND SWITCHES

For most local indicators (including pressure), installation costs must be as low as possible. If an indicator requires an external source of power (no matter how low), one faces wiring costs, including protection appropriate to the environment in which the wiring is installed. Therefore, indicators that require no external source of power are always less expensive to install. For pressure indicators, pressure gauges are the norm. While microprocessor-based versions with digital displays are available, few are installed in industrial plants due to the wiring issues. Battery-powered versions may eventually change this practice, but at the moment batteries are generally avoided in industrial environments.

Switches are used to detect when a variable exceeds a specified value. Their output is a discrete value. The output for pressure switches is usually considered to be either high/normal or normal/low. The inputs to most shutdown and safety systems are discrete values from such switches. For these applications, reliability is crucial. The way you get good reliability is to keep it simple. Consequently, most pressure switches are entirely mechanical in nature and thus require no external source of power. Electronic pressure switches are on the market. However, the simplicity of mechanical switches is especially attractive for use in safety systems.

Pressure Gauges

A pressure gauge is a pressure indicator. Most couple a Bourdon tube of some type to a circular indicating dial via a suitable mechanical mechanism. A few gauges, especially those sensing pressure of a few inches of water, are constructed from bellows. Factors to consider in selecting a pressure gauge include

Pressure range. For sensing gauge pressure, the measurement range is from zero to a specified upper-range value. The operating pressure should not exceed 75% of the upper-range value. Excess pressures can permanently damage a Bourdon tube and even cause it to burst. Mechanical stops in some models provide additional protection from excess pressures.

Temperature range. A typical ambient temperature range for a pressure gauge is −40° to 70°C. The process temperature may be higher provided the gauge is mounted sufficiently far away so the temperature of the gauge is in the acceptable range. The fluid inside the gauge must also be considered. Specifically, freezing temperatures in winter may necessitate heat tracing.

Materials of construction. The wetted parts of gauges for noncorrosive services are usually constructed from either brass (for low pressures) or stainless steel. Most gauge manufacturers provide numerous alternatives for corrosive applications.

Vibrations and pulsations. Excessive mechanical vibrations from the process equipment and pressure pulsations from within the process must be avoided.

Pressure Switches

A pressure switch changes state at a pressure above atmospheric pressure. A vacuum switch changes state at a pressure below atmospheric pressure. A differential pressure switch changes state based on a differential pressure. The mechanisms are basically the same, so hereafter we shall discuss only pressure switches.

A pressure switch consists of two components:

Pressure-sensing element. Most use a diaphragm, but a few use a bellows. Bourdon tubes are rarely used in pressure switches.

Electrical-switching element. Most couple a snap-action electrical switch to the pressure-sensing element.

The electrical-switching element becomes a component of a circuit whose behavior depends on the state of the switch. The circuit for a PLC generally consists of a source of power, the switch, and the PLC's input module. This permits the PLC to sense the state of the switch based on the circuit's continuity or lack thereof.

3.5. PRESSURE SENSORS

An electronic pressure transducer essentially consists of two elements:

Pressure-sensing element. This is usually a diaphragm that converts pressure into a displacement. Bourdon tubes and bellows are rarely used.

Displacement-sensing element. This converts the displacement from the pressure-sensing element into an electrical property (voltage, resistance, capacitance, etc) that can be sensed.

The three most commonly encountered technologies for the displacement-sensing element are the following:

• Strain gauge.
• Capacitance.
• Resonant frequency.

There are other choices, usually more mechanical with less accuracy but less expensive. For example, some pressure gauges are able to output a current or voltage signal that represents the pressure. In addition to moving the dial pointer, the mechanism attached to the Bourdon tube also moves the wiper arm of a precision potentiometer. This gives a variable resistance that is proportional to the pressure. While inexpensive, the accuracy is not even close to what can be achieved with electronic pressure sensors.

Measurement Ranges

Before the introduction of the electronic pressure transmitters, the user had to specify the required measurement range at the time the transmitter was ordered. Although the upper-range value could be subsequently adjusted, the adjustment range was limited.

With electronic pressure transmitters, the manufacturers are able to reduce the number of different models of pressure sensors that are required to cover the measurement ranges needed by most users. For example, one approach is to provide a low-pressure model, a medium-pressure model, and a high-pressure model with the following capabilities for differential pressures:

Model	Span Limits (in H_2O)	Range Limits (in H_2O)
Low pressure	2.0 to 40	−40 to 40
Medium pressure	4.0 to 400	−400 to 400
High pressure	20 to +2000	−2000 to +2000

These specifications also reflect another aspect that is characteristic of the electronic pressure transmitters: The high side and low side are at the discretion of the user. Electronic pressure sensors are inherently bidirectional; reversing the high-side connection with the low-side connection merely gives a differential pressure of the opposite sign. This is not the case with mechanical pressure sensors.

Also note the overlap in the ranges. With this flexibility, most users could meet their requirements with two models, and a few could do so with only one. At least one manufacturer uses only two models of pressure sensors (a low-pressure version and a high-pressure version whose ranges overlap).

Suppliers provide a large number of pressure-transmitter models. There are different models for gauge pressure, absolute pressure, and differential pressure. All electronic pressure transmitters use isolating diaphragms to protect the internals from process fluids, which significantly increases the number of models. This in turn increases the incentives for reducing the number of pressure-sensor models.

For the supplier, reducing the number of models simplifies manufacturing, inventory, etc. For the user, there are two distinct advantages:

- When ordering the pressure transmitters required for a new plant, it is crucial to order only the model with the appropriate pressure sensor.

After the transmitter is installed, the required measurement range can be specified when the transmitter is configured. If the initial measurement range proves to be unsatisfactory, it can be easily changed.

- All plants either maintain a stock of spare parts or contract with an outside company to do so. By reducing the different models, the spare parts inventory is simplified and reduced.

Performance

The performance parameters for the electronic pressure transmitters vary somewhat with the type of pressure sensor, but mostly at the fringes. The following are the lowest common denominators of the performance parameters:

Pressures	Gauge, absolute, and differential
Accuracy	0.1% of span
Rangeability	100:1
Measurement range	0.5 to 4000 psig for gauge pressure 2.0 to 1000 in H_2O for differential pressure
Materials	Wide selection for isolating diaphragms
Drift	0.02% of upper-range limit per year
Temperature	$-25°$ to $85°C$

These performance parameters meet the requirements for most users. Very few users have requirements on the edge of the envelope, so most users would not base their selection on the performance parameters resulting from the type of pressure sensor within the transmitter. Price is also comparable and is usually negotiable if you are ordering a large number of transmitters. Often the path of least resistance is simply to order the same model that was ordered the last time, although occasionally factors such as the availability of local support or spare parts influences the decision.

3.6. STRAIN GAUGE PRESSURE SENSORS

For purposes herein, a strain gauge is a component whose resistance changes as it is stretched (in tension) or compressed. Early strain gauges were thin metal wires; more recent strain gauges are silicon and fabricated using semiconductor technology. A strain gauge pressure sensor is constructed as follows:

Diaphragm. The process pressure causes the diaphragm to deform.

Strain gauge. The deformation of the diaphragm is sensed by one or more strain gauges. In most configurations, two (or a multiple of two) strain gauges are attached to the pressure-sensitive diaphragm. As the diaphragm deforms, half of the strain gauges are in tension (resistance increases) and the others are in compression (resistance decreases).

Resistance sensor. One way to sense the resistance of the strain gauge is to flow a small current through the gauge and measure the voltage across it. But when two or more strain gauges are used, bridge arrangements, such as the Wheatstone bridge, are superior for sensing the difference between the resistances of the strain gauges in tension and the strain gauges in compression.

Strain in Diaphragms

As used in pressure transmitters, diaphragms are circular disks clamped at the edges. There are two types of strain:

Tangential strain ε_T. This strain is maximum at the center and zero at the edges.

Radial strain ε_R. At the center, the radial strain is the same as the tangential strain. But the maximum radial strain is at the edges, being twice the strain at the center but opposite in sign (in compression instead of tension).

Figure 3.8 illustrates, each strain varies in a parabolic fashion with distance from the center. For small deformations, the radial strain is expressed by the following equation:

$$\varepsilon_R = \frac{3PR^2(1-v^2)}{8t^2E}$$

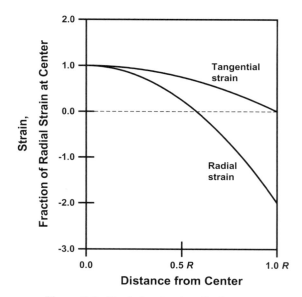

Figure 3.8. Strain in circular diaphragms.

where ε_R = radial strain at center; P = pressure (psi); R = radius of diaphragm (in); ν = Poison's ratio; t = diaphragm thickness (in); E = Modulus of elasticity (psi).

Similar equations are available for the deflection at the center and the resonant frequency. These relationships permit the strain gauges to be precisely positioned such that the pressure can be most accurately determined from the strain sensed by the strain gauges.

Thin Film

In early versions of strain gauge pressure transmitters, metallic strain gauges were bonded (glued) to the surface of the pressure sensitive diaphragm. Less than satisfactory experience with adhesives led to the *thin-film pressure sensor:*

- A thin film of nonconductive material such as silicon is deposited on the pressure sensitive diaphragm.
- The metallic material for the strain gauge is deposited on the nonconductive film by a sputtering process, which creates a molecular bond instead of an adhesive bond. A mask applied to the nonconductive film determines the location of the strain gauges.

For both performance and reliability, thin-film pressure sensors are quite acceptable. However, the manufacturing process is tedious and costly.

Piezoresistivity

Monocrystalline silicon provides very high mechanical strength with little hysteresis and only minor mechanical changes due to overload or aging. A monocrystalline silicon wafer can be micromachined to form a very small pressure sensitive diaphragm.

Monocrystalline silicon that is doped with boron exhibits piezoresistivity— that is, its resistance changes with strain. By using a diffusing process, semiconductor strain gauges in the form of a Wheatstone bridge are incorporated directly into the pressure-sensitive diaphragm. The result is a microelectromechanical system (MEMS). The usual integrated circuit technology is two-dimensional; MEMS uses the third dimension to create sensors that are both very small and very inexpensive ($10?). Because the piezoresistive pressure sensor is an IC, extensive logic (even an A/D converter) can be incorporated during the manufacturing process. We may eventually see all of the transmitter logic implemented within the pressure sensor itself.

Figure 3.9 depicts a differential pressure transmitter that uses a piezoresistive sensor. So that the sensor is not exposed to process fluids, isolating diaphragms and oil are used to transmit the external pressures to the sensor. Gauge and absolute pressure configurations are available.

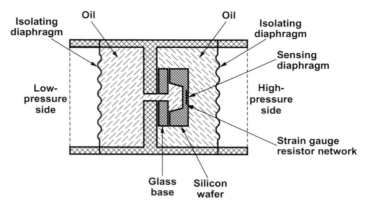

Figure 3.9. Piezoresistive pressure sensor.

Overpressure Protection

Using semiconductor technology to fabricate the pressure sensor has a characteristic that is both an advantage and a disadvantage.

The recovery of the sensor to a pressure in excess of its rating is excellent with one disclaimer—that the sensor does not rupture. That is, if the sensor does not rupture, it will continue to function properly once the pressure returns to within the rated range. However, should it rupture, it has the potential of fluid leaking through the sensor.

The design of the electronic pressure transmitters shown in Figure 3.9 is modified so that it is not possible for the internal filling oils to transmit a higher pressure than the pressure sensor can withstand. One approach is to limit the possible displacement of the isolating diaphragms. However, this is more detailed than is of interest to most users, so it will not be pursued herein.

3.7. CAPACITANCE PRESSURE SENSORS

A capacitor consists of two metal plates separated by a dielectric material (a material that can store and release electrons). As current flows into a capacitor, the voltage across the capacitor increases. This is described by the following equation:

$$C = \frac{dV}{dt} = I$$

where C = capacitance; I = current; V = voltage.

The larger the capacitance C, the more slowly the voltage changes for a given current flow. The value of the capacitance C depends on the following:

- Dielectric constant of the material between the plates of the capacitor. Capacitance increases with dielectric constant.
- Distance D between the plates. Capacitance increases with D.

To sense pressure, a capacitor must be designed so that the distance D changes with the process pressure.

Sensing the Capacitance

A capacitor is basically a dynamic element. If a DC voltage is applied across the plates, the current flows until the voltage across the capacitor is equal to the DC voltage being applied. The dynamic behavior is that of a first-order lag. Once equilibrium is attained, the voltage is constant and the current flow is zero.

To sense the capacitance C, a change must be occurring—that is, the capacitor voltage must be changing so that the current flow will not be zero. The simplest way to do this is to use an AC voltage instead of a DC voltage. Now current alternately flows into the capacitor and then out of the capacitor on each cycle in the voltage. As Figure 3.10 illustrates, this requires the following:

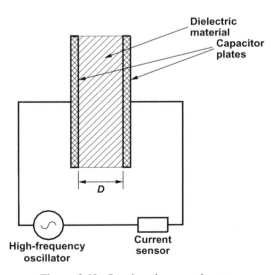

Figure 3.10. Sensing the capacitance.

High-frequency oscillator. This is essentially an AC voltage source.

Current sensor. This senses the AC current flowing into and out of the capacitor. This current flow is proportional to the capacitance of the capacitor.

Diaphragm

As used in industrial pressure transmitters, the pressure-sensing diaphragm constructs two capacitors. By inserting the metal diaphragm between two fixed metal plates (Fig. 3.11), two capacitors are created:

- One capacitor is between the diaphragm and the left fixed metal plate.
- One capacitor is between the diaphragm and the right fixed metal plate.

A high-frequency oscillator and current sensor are required for each capacitor. If the diaphragm is exactly in the middle, the two capacitances are equal and the two currents are also equal.

Suppose the pressure is applied such that the diaphragm moves slightly to the right. The capacitance of the capacitor on the left increases; the capacitance of the capacitor on the right decreases. The current sensor on the left senses an increase in the current flow; the current sensor on the right senses a decrease in the current flow. The difference in the current flow reflects the difference in the capacitances, which in turn reflects the deformation of the diaphragm.

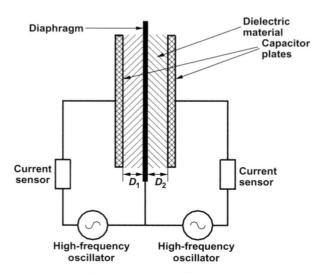

Figure 3.11. Pressure-sensitive diaphragm.

Capacitance Cell

The practical implementation to create two capacitors with a pressure-sensitive metal diaphragm is illustrated in Figure 3.12. The pressure-sensitive diaphragm is in the center of the cell and is normally grounded to the pressure transmitter's case.

There are two other diaphragms, one on each side of the cell. These are isolating diaphragms that retain the dielectric material (an oil) and isolate the cell internals from the process fluids. The isolating diaphragms do not offer any resistance to a change in pressure but merely transfer the exterior pressure to the oil inside the cell.

The pressure-sensitive diaphragm is one plate of each capacitor. Within each cavity, a metal plate is inserted to provide the other plate of a capacitor. The high-frequency oscillator and current sensors provide the same functions as previously described. As the pressure on the left increases relative to the pressure on the right, the pressure-sensitive diaphragm moves slightly to the right. This increases the capacitance on the left and decreases the capacitance on the right. In turn, this increases the current flow on the left and decreases the current flow on the right. This difference in current flow can be related to the difference in pressure across the cell.

Figure 3.12 suggests two high-frequency oscillators, but in practice only one is required. Furthermore, current sensors are not the only option for sensing the difference in capacitance. Another possibility is a capacitor bridge network, which is analogous to the Wheatstone bridge network for resistors.

The capacitance cell pressure transmitter appeared during the transition from pneumatic to electronic instrumentation. With performance far superior

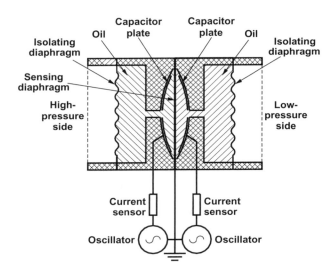

Figure 3.12. Capacitance cell for measuring pressure.

to its predecessors, it also proved to be very robust and reliable. The early strain gauge pressure sensors relied on bonding the strain gauges to the pressure-sensitive diaphragm, which ultimately proved to be unacceptable.

In the 1980s, the capacitance cell dominated the market and continues to do well in the very competitive market for pressure transmitters. However, it is too costly to effectively compete in mass markets such as pressure sensors for use in automobile engines. Advancements in semiconductor technology have made the piezoresistive strain gauge a formidable competitor, especially in applications in which the sensor can be directly exposed (no isolating diaphragms and filling oil).

3.8. RESONANT FREQUENCY

Resonance is a forced vibration. When a mechanical system is excited, it vibrates or resonates at a frequency called the resonant frequency. When the external excitation is also at the resonant frequency, the maximum vibration is exhibited by the mechanical system. At the resonant frequency, very low power is required to induce measurable vibrations into the mechanical system.

Two mechanical designs for pressure sensors based on the resonant frequency are possible:

Resonant wire. A thin wire is fixed at one end and attached to a pressure-sensitive diaphragm at the other. A change in pressure causes the deformation of the diaphragm to change. This affects the tension in the wire, which affects its resonant frequency.

Resonant cylinder. A thin cylinder has vacuum on the outside and process gas on the inside. A change in process pressure changes the force on the cylinder, which changes its resonant frequency.

Electromagnetic drivers excite the mechanical element and the frequency at which it vibrates is sensed. The resonant frequency is obtained by adjusting the frequency of the excitation to match the frequency at which the element is vibrating. The process pressure is then inferred from the resonant frequency and the characteristics of the mechanical element.

Resonant frequency also occurs in electrical systems. The resonant frequency for an electrical system is much higher than the resonant frequency for a mechanical system. For use in process pressure measurements, electrical systems have proven to be far superior to their mechanical counterparts.

Silicon Resonant Sensor

A single silicon crystal is machined to give a pressure-sensitive diaphragm of the appropriate geometry and thickness. Two resonators are positioned on the pressure-sensitive diaphragm. Deformation of the diaphragm causes strain in each of the resonators. The strain on one resonator is tension, which causes the resonant frequency to increase. The strain on the other resonator is compression, which causes the resonant frequency to decrease. The difference in the resonant frequencies of the two resonators is sensed. If the geometry of the diaphragm and the characteristics of the resonators are known, the pressure can be inferred from the difference in the resonant frequencies.

Like all semiconductor products, these sensors are very small, so magnified photos are required to reveal any details. But unless you are quite familiar with semiconductor technology (and most of us are not), they look impressive but otherwise do not mean very much. Yokagawa pioneered this technology, so those interested in more details on the resonant frequency pressure sensor can retrieve more information from Yokagawa's website.[1]

Like the piezoresistive pressure sensor, the silicon resonant pressure sensor is also a MEMS. Products based on this technology are highly stable. They exhibit excellent repeatability, with almost no hysteresis. Their characteristics are not significantly affected by aging.

For use within industrial pressure transmitters, isolating diaphragms and filling oils are required in much the same manner as for the piezoresistive sensor illustrated in Figure 3.9. The overpressure issues are also very similar, so they will not be repeated.

Why Resonant Frequency

Basic sensor technology can be divided into two categories:

Analog sensors. The basic sensor output is a voltage signal (or is converted to a voltage signal) that is then converted to digital using a D/A converter. Both the piezoresistive pressure sensor and the capacitance cell are analog sensors.

Digital sensors. The basic sensor output is a frequency signal, actually a binary signal that changes state at a frequency that depends on the variable being sensed. A microprocessor can sense a frequency directly, so no D/A converter is required. The resonant frequency pressure sensor is a digital sensor whose output frequency is 100 kHz, more or less.

With analog sensors, the precision of the measurement is ultimately limited by the resolution of the D/A converter. In an industrial environment, 12 data bits (1 part in 4000) is practical. But going beyond 12 requires superb signal conditioning, without which the additional bits are largely noise. The appeal of digital sensors is that there is no limit to the resolution that can be achieved.

Frequency Input

A microprocessor can determine the frequency (Hz or cycles/second) using either of the following:

Timer. Determine the time from this 0 to 1 transition to the next 0 to 1 transition. This time is the period of the cycle; the frequency (Hz) is the reciprocal of the time (sec). The accuracy depends on the resolution with which the microprocessor senses the time interval between the 0 to 1 transitions. This approach is normally used only for low-frequency inputs (<100 Hz).

Counter + Timer. Determine the time for the input to make a specified number of 0 to 1 transitions or, conversely, count the number of 0 to 1 transitions in a specified interval of time. The frequency (Hz) is the count divided by the time (sec). This approach works best for high-frequency inputs; the higher the frequency, the better (unless the frequency is so high that the microprocessor misses some of the 0 to 1 transitions).

The appeal of the latter is that there is no limit on the resolution that can be achieved. As noted earlier, the resonant frequency pressure sensor is a digital sensor whose output frequency is 100 kHz, more or less. At 100 kHz, the period of the cycle is 10 msec. Suppose we count the number of pulses in 10 msec, which permits the measured value to be updated 100 times per second. The resolution is the count ±1. Suppose we change the time interval to 100 msec. The count will increase by a factor of 10, but the resolution is still the count ±1. Basically, the resolution for the measured value has improved by a factor of 10. There is a downside: The measured value can be updated only 10 times per second. But for process applications, relatively slow update rates can be tolerated.

The analog approach is capable of meeting today's requirements for accuracy and resolution in industrial pressure transmitters. But as always, the requirements continue to become more demanding. This is basically the case for digital sensors—at some point, our requirements will exceed what is possible with A/D converters. That this will occur appears inevitable; when it will occur is difficult to project.

3.9. INSTALLATION

The installation of pressure transmitters and pressure gauges differs depending on the nature of the fluid:

- Liquid.
- Gas. This arrangement can also be used for condensing vapors for which the vapor temperature is within the temperature limits of the pressure transmitter (typically −25° to 85°C).
- Condensing vapor. This arrangement is designed to prevent the pressure transmitter from being directly exposed to the vapor and must be used when the vapor temperature exceeds the temperature limits for the pressure transmitter. For example, steam condenses at temperatures above 100°C, which is above the temperature limits for the pressure transmitter.

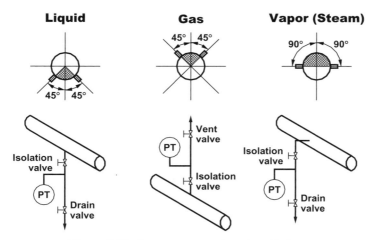

Figure 3.13. Pressure transmitter installation.

Figure 3.13 illustrates the recommended installation for each. Most companies have a standard installation for each case, which may differ slightly from those in the figure. However, the issues to be addressed are basically the same.

To permit servicing without disrupting process operations, the setups shown in Figure 3.13 provide an isolation valve and either a drain valve or a vent valve. Some installations must also provide for calibration. Most suppliers of pressure transmitters also supply a manifold that provides two, three, or four connections to facilitate such functions. Machined from a single block of metal, the reduction in the number of connections reduces the likelihood of leaks and other problems. Most manufacturers will supply the pressure transmitter connected to the required manifold.

Liquid Pressure

The installation must be such that the connection lines from the transmitter to the process are completely full of liquid. For installations on pipes, this is achieved as follows:

- The tap should be within 45° of the bottom of the pipe. However, if sediments are present, the tap should be at the 45° angle instead of at the bottom of the pipe.
- The connection piping should be arranged such that any gases that may form within the connection piping will escape into the main pipe.
- The pressure transmitter should be positioned below the tapping point and arranged such that any gases within the transmitter can escape upward.

Gas Pressure

The installation must not allow liquid to accumulate in the connection lines from the transmitter to the process. For installations on pipes, this is achieved as follows:

- The tap should be within 45° of the top of the pipe.
- The connection piping should be arranged such that any liquid that may form within the connection piping will drain into the main pipe.
- The pressure transmitter should be positioned above the tapping point and arranged such that any liquid within the transmitter can drain downward.

Condensing Vapor

The installation must be such that the connection lines from the transmitter to the process are completely full of condensate. For installations on pipes, this is achieved as follows:

- The tap should be at or above the horizontal centerline of the pipe.
- The connection piping should be arranged so that it will completely fill with condensate. Any noncondensable gases that may form within the connection piping must escape into the main pipe.
- The pressure transmitter should be positioned below the tapping point and arranged so that it will be completely full of condensate. Any non-condensable gases within the transmitter must escape upward.

The temperature of live steam is above the temperature rating for electronic pressure transmitters. The connection piping must provide sufficient distance between the transmitter and the pipe so the condensate will be adequately cooled. Unfortunately, in cold weather, the condensate may freeze, so appropriate tracing will be required. It is also necessary to fill the connection piping with condensate before start-up.

Siphon Loop

In pressure gauge installations on hot, condensing vapors such as steam, a siphon loop, as illustrated in Figure 3.14, is often installed to prevent the gauge from being exposed to the hot vapors.

Before start-up, the siphon loop must be filled with condensate. During service, it will remain filled to the top of the siphon loop. During operation, condensate continues to form in the connection piping, and thus any loss of condensate from the siphon loop will be replenished.

Siphon loops are common for pressure gauge installations, but not for pressure transmitter installations.

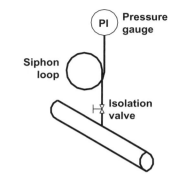

Figure 3.14. Siphon loop.

Figure 3.15. Purged connection.

Purging

An occasional pressure transmitter installation encounters problems such as the following:

- Buildups or sediments plug the connection piping.
- The transmitter must not be exposed to the fluids in the process.
- The pressure measurement point is in an inaccessible location.

One way to address these problems is via purging, using arrangements such as those illustrated in Figure 3.15. For gas or vapor applications, the purge fluid is a gas. For liquid applications, the purge fluid may be either a gas or a liquid. For liquid applications, the transmitter should be at the same elevation (or just slightly below) as the process tap. If the difference in elevation is significant, the liquid head must be subtracted from the pressure measurement (normally by elevating the transmitter's zero).

Cost of purging is rarely an issue. The volume of the purge fluid is small and an inexpensive fluid can be used. However, an adequate purge flow rate must be maintained at all times. Usually the flow is indicated locally via a rotameter. The plant operators are responsible for checking the flow on a regular basis. However, things can go wrong, and in industrial plants, they eventually will. The result is plugged connection piping, a transmitter exposed to corrosive fluids, etc.

Direct Mount (Flange Mount)

Where buildups or sediments cause problems in the connection piping, direct-mount or flange-mount pressure transmitters can be considered. These are most often encountered in level measurement applications in which the variable actually being sensed is the pressure at the location of the transmitter.

In some applications, it is necessary to eliminate any dead space, where the materials may not be representative of the remaining contents of a vessel. Especially when a chemical reaction is occurring in a vessel, the composition of the dead space may be quite different, leading to undesirable reactions occurring there. Such versions are referred to as flush-mount transmitters.

Direct mount eliminates the connection piping that was shown in the earlier figures. However, there is one down side to a direct-mount installation: The direct-mount transmitter cannot be removed without shutting down the part of the process in which it is installed.

Snubber

Significant levels of pressure pulsation or vibration will lead to noise on the measured pressure. For pressure gauge installations, they will quickly have an adverse affect on the gauge itself.

In these situations, a porous metal snubber is often incorporated into the connection piping, as illustrated in Figure 3.16. Snubbers can be used for either liquid service or gas service, but they are not interchangeable. The snubber dampens (attenuates) the pressure pulsations at the pressure transmitter. However, the term *pulsation damper* is generally understood to be a device that is installed on the discharge of a positive displacement pump to dampen both flow and pressure pulsations.

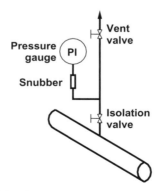

Figure 3.16. Snubber in a gas pressure measurement application.

The snubber affects only the dynamic behavior of the pressure indication. The contribution of the snubber is the same as that of a first-order lag inserted between the process and the pressure measurement. A snubber is not appropriate where the objective is to detect a rapid change in the process pressure.

Snubbers are recommended in pressure gauge installations only because of the adverse affect of pressure pulsations on the mechanical components of the pressure gauge. Snubbers are not recommended for electronic pressure transmitter installations.

3.10. DIFFERENTIAL PRESSURE

Measuring the difference in pressure between two points in a process often requires a significant length of piping for at least one of the connections. When this piping is filled with liquid, the differential pressure sensed by the transmitter may have a hydrostatic head component. This raises a number of issues.

Column Pressure Drop

Distillation towers, strippers, and absorbers use either packing or trays for vapor–liquid contact. The distillation tower illustrated in Figure 3.17 contains packing, but the issues are the same for trays. In most towers, differential pres-

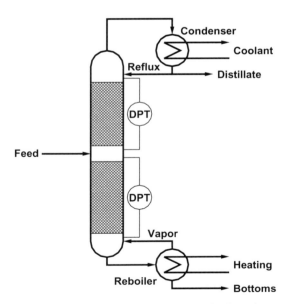

Figure 3.17. Measuring pressure drop across packed sections of a tower.

sure transmitters are installed to measure the pressure drop across each section. But because the issues are the same for each section, we will subsequently examine only the lower packed section.

Physical access to equipment, including instrumentation, is an issue for some towers:

- The large towers in refineries and commodity chemical plants have sufficient structural strength to stand on their own. If any instrumentation is physically located at the feed location or at the top of the tower, someone has to climb the tower to get physical access to it. To avoid a large weight at the top of the tower, the overhead vapor is piped down the tower to a condenser at ground level, and the condensate (reflux) is pumped back up the tower. This permits some of the instrumentation to be physically located at grade level.

- For specialty chemicals, the tower diameters are much smaller. For structural support, a multistory structure must be built around the tower. This structure also allows physical access to instrumentation and other equipment. With a structure available for support, equipment such as condensers and reflux drums can be physically located at the top of the tower.

Dry Legs

One approach is to physically locate the differential pressure transmitter slightly above the packed section, as illustrated in Figure 3.18. With this con-

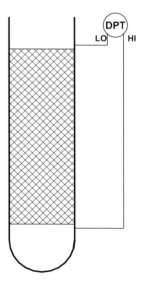

Figure 3.18. Dry legs.

figuration, both of the connecting lines (or *legs*) to the differential pressure transmitter contain vapor, hence the term *dry legs*. The following observations apply to this configuration:

- This arrangement is not consistent with the recommendations for a condensing vapor—that is, that the connecting lines should be filled with condensate. One consequence of this is that the transmitter will be exposed to the hot vapors from the tower. Especially for the lower packed section, these temperatures are likely to be above the temperature limits for the differential pressure transmitter.
- The differential pressure transmitter is physically located at or just above the feed tray. For large towers that stand on their own, this is not an easily accessible location.

Wet Legs

In the previous discussion on installing pressure transmitters for condensing vapors, the recommendation was that the connection lines should be filled with condensate. To do this, the differential pressure transmitter must be physically located below the packed section, as illustrated in Figure 3.19. With this configuration, both of the connecting lines to the differential pressure transmitter contain condensate, hence the term *wet legs*. The following observations apply to this configuration:

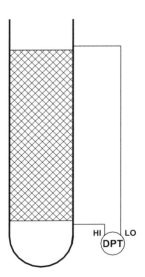

Figure 3.19. Wet legs.

- The connection piping must be completely filled with condensate. That means that all lines must be sloped so that any gases or vapors that form in the connection piping can escape upward into the column.
- The transmitter is now exposed to condensate. The temperatures at the bottom of the tower are higher than at the middle or top, but sufficient lengths of connection piping can be provided so that the temperature at the transmitter is acceptable.
- The transmitter can be located at grade level for easy physical access.
- The feed to a tower is always a mixture. The lights (components with low boiling points) concentrate at the top; the heavies (components with high boiling points) concentrate at the bottom. What is the composition of the condensates in the wet legs? Hard to say. But this is an important question because density is a function of composition.

HI and LO

The process pressure at the lower tap to the column is higher than the process pressure at the upper tap. But with wet legs, the differential pressure at the transmitter is not the same as the process differential pressure. For the wet leg connected to the upper tap, the pressure at the transmitter is the process pressure at the upper tap plus the liquid head of the wet leg.

The conventional pressure transmitters could sense only positive differential pressures—that is, when the pressure at the HI connection is greater than the pressure at the LO connection. The electronic pressure transmitters can sense both positive and negative pressures, which makes the HI and LO connections somewhat arbitrary in that they can function in either configuration.

The options for connecting the transmitter to the tower taps are illustrated in Figure 3.20:

Lower tap to HI, upper tap to LO. When there is no vapor flow through the tower, the differential pressure at the transmitter will be the liquid head of the wet leg but will be negative. The output of the transmitter is to be zero for this differential pressure. As the vapor flow though the tower increases, the differential pressure sensed by the transmitter increases (actually becomes less negative). The output of the transmitter will increase as the differential pressure increases.

Lower tap to LO, upper tap to HI. When there is no vapor flow through the tower, the differential pressure at the transmitter will be the liquid head of the wet leg. The output of the transmitter is to be zero for this differential pressure. As the vapor flow though the tower increases, the differential pressure sensed by the transmitter decreases. But smart

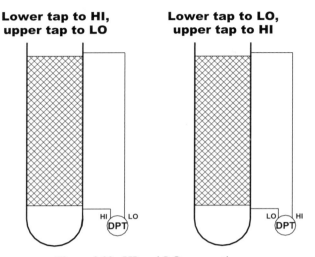

Figure 3.20. HI and LO connections.

transmitters can be configured so that the transmitter output increases as the sensed differential pressure decreases.

With smart transmitters, the procedure for configuring the differential pressure transmitter is basically as follows:

Specify one point on the transmitter operating line. For the tower differential pressure application, the differential pressure is sensed by the transmitter under the following conditions:
  • There is no vapor flow in the tower.
  • The legs are filled as in normal operations.
For the differential pressure sensed under these conditions, specify the value to be indicated for the process differential pressure.

Specify the span. For the measured value, this will be the span in process differential pressure units.

The configuration procedure can be applied for wet legs or dry legs. In current loop applications, it is also possible to "reverse" the transmitter output—that is, output 20 ma for a process differential pressure of zero and then decrease the transmitter output as the process differential pressure increases. On failure of the current loop, the indicated differential pressure will be the upper-range value instead of the lower-range value (normally zero for process differential pressures).

Purging

In tower applications, condensate in a wet leg can be a problem. Its composition is always a question. For the tap at the bottom of the tower, the condensate should be close to that of the bottom product. In some applications, this is a quite viscous material at the temperatures acceptable at the differential pressure transmitter.

Figure 3.21 illustrates purging both legs. When purging with a gas, the net result is dry legs. While it is possible to purge only one of the legs, usually both are purged. The issues are

- The problems with purging were previously discussed, but the main problem is that the purge must be maintained at all times, which proves to be more difficult than it would appear.
- The transmitter is not exposed to the materials within the tower or to tower temperatures.
- The transmitter can be located at any convenient physically accessible location.

Capillary Seal Systems

Most differential pressure transmitter manufacturers will provide diaphragm seal systems for the low and/or high sides of their differential pressure transmitters. A capillary seal system consists of two components:

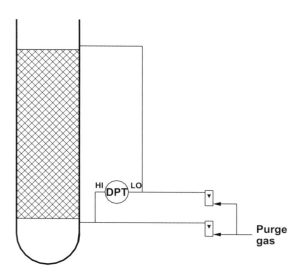

Figure 3.21. Purged connections.

- A diaphragm seal, usually in a flanged process connection, translates the process pressure to the filling fluid. Most manufacturers offer a variety of filling fluids for different temperature ranges and to meet sanitary requirements, if applicable.
- Flexible capillary tubes convey the pressure to the differential pressure transmitter. Lengths up to about 15 m are standard, with longer lengths for custom orders.

Capillary seal systems cannot be homemade. Filling must be done with proper equipment and by knowledgeable technicians. The seal system must be completely filled with liquid (no gas pockets, and no dissolved gas in the liquid that can subsequently form gas pockets). The seal system must not be overfilled, thus creating its own pressure. Every plant has its tinkerers who will take anything apart, but capillary seal systems must not be compromised. If so, they must be returned to the manufacturer (or other qualified facility) to be refilled.

Use on Tower

The use of capillary seals for the tower pressure drop measurement is illustrated in Figure 3.22. The result is a wet leg configuration, but now the contents of the wet legs are known.

With no vapor flowing through the tower, the differential pressure at the pressure transmitter will be a liquid head of a height of the filling material equal to the difference between the tap locations. With the upper tap connected to the LO side, the differential pressure at the transmitter will be

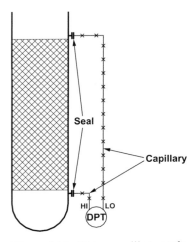

Figure 3.22. Using capillary seals.

negative. Using either zero suppression or its equivalent, the transmitter Is configured to output a value of zero. As the vapor flow increases, the differential pressure increases (actually becomes less negative), and the output of the transmitter increases.

With capillary seal systems, the pressure transmitter can be physically located wherever convenient. However, physical access to the diaphragm seals is possible only when the tower is not in operation.

Two Pressure Transmitters

One way to avoid the issues associated with wet legs in differential pressure measurements is to measure each pressure individually and then compute the difference. This is illustrated in Figure 3.23 for the tower. This certainly simplifies the connecting piping as well as avoiding purging and capillary seals. However, this occurs when the pressure transmitter at the upper tap is physically located near the tap. Unfortunately, this location is not easily accessible for the large towers.

Computing the pressure difference from two pressures can lead to subtracting two large numbers to obtain a small number, which is not a good numerical practice. For a depropanizer, the tower pressure is 300 psig, more or less. Suppose the pressure drop across a section is about 5 psi. Let's use a smart pressure transmitter with an accuracy of ±0.1% of its upper-range value. If the tower pressure transmitter upper range-values are 400 psig, this means an error of ±0.4 psig. In a measurement of 5 psi, this is 8%!

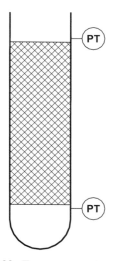

Figure 3.23. Two pressure transmitters.

The numerical issues suggest that this approach is more acceptable in low-pressure towers than in high-pressure ones. Errors in either transmitter's zero adjustment also contribute to the error in the difference. Because their zero errors are much smaller than their predecessors, smart transmitters are recommended.

LITERATURE CITED

1. The URL to Yokogawa's web site for resonant frequency information is as follows: http://www.yokogawa.com/fld/reference/fld-si-r-01en.htm

CHAPTER 4

Level and Density

This module examines various technologies for measuring level (and interface level):

Pressure (and differential pressure). About 75% of the level measurement applications rely on pressure or differential pressure. This technology is generally well known and economical, but there are limitations, such as not being applicable to solids.

Capacitance. This technology can be applied to solids as well as liquids. It is invasive, but usually the probe can be installed through a single opening in the top of the vessel.

Ultrasonic. Also applicable to solids and liquids, this technology is noncontact, but does protrude through a single opening in the top of the vessel.

Noncontact radar. This approach relies on the same principles as the familiar radio direction and ranging (radar) to provide the most accurate level measurement technology currently available.

Guided wave radar. Especially when applied to small vessels, the noncontact radar systems must cope with multiple reflections. Guided wave radar uses a wave guide to avoid these issues, but at the expense of being an invasive technology with a probe comparable to that used in capacitance level measurement devices.

Nuclear. This technology is noncontact and noninvasive, with all equipment mounted external to the vessel. Despite giving good performance even in extremely difficult applications, the presence of radioactive materials on site generally makes this technology the measure of last resort.

Others. This includes displacers, resistance tapes, and a variety of float systems.

Because many of these can also be applied to measure density, measurement of density (or specific gravity) is also covered.

Basic Process Measurements, by Cecil L. Smith
Copyright © 2009 by John Wiley & Sons, Inc.

4.1. LEVEL, VOLUME, AND WEIGHT

Level is a measure of the physical location of an interface. In most industrial applications, the interface is the surface of a liquid in a tank. However, level can be applied to any of the following interfaces:

Lower Layer	Upper Layer
Liquid	Gas
Liquid	Liquid
Solid	Gas
Solid	Liquid

Level is a length and has units such as meters and feet. If one is concerned about overflowing a vessel, the physical location of the liquid surface is of paramount interest. But for other applications, a level measurement becomes the basis for expressing the contents of the vessel. For a vessel containing liquid, the level measurement can be translated to the following:

Volume of liquid in the vessel. The translation of level to volume depends on the physical geometry of the vessel.

Mass of liquid in the vessel. One first translates from level to volume and then multiplies by the liquid density to obtain the mass of liquid.

Percent capacity. In this context, percent may be based on the height of liquid in the tank or on the volume of liquid in the tank. For straight-walled vessels (constant cross-sectional area), these are equivalent.

Vessel Geometries

There are many possible variations in vessel geometry. The most common are the cylindrical and spherical geometries illustrated in Figure 4.1. The cylinder may be either horizontal or vertical. Cylindrical vessels may have ends of various shapes, the most common being flat, dished, or conical.

The options for converting between level and volume are

Formula. The simplest is a vertical cylinder with a flat bottom. It can also be easily applied to a dished bottom provided the level is always in the straight-walled section. Formulas can be developed for many other vessel geometries. However, the equations will be nonlinear. Especially if one has to compute level from volume, analytic expressions may not be possible.

Strapping table. This is normally a table or graph that provides the volume as a function of level. For complex vessel geometries and when vessel internals such as coils must be considered, this is the only practical

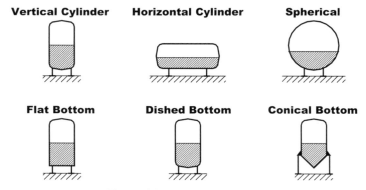

Figure 4.1. Vessel geometries.

approach. Normally the vessel manufacturer supplies the data. Within digital systems, the strapping table can be represented by either a table lookup function or a characterization function.

Density and Specific Gravity

The relationships among density, specific volume, specific gravity, and a few special measures are summarized as follows:

Measure	Symbol	Relationship	Units
Density	ρ	Mass per unit volume	g/cc or lb_m/ft^3
Specific volume	v	Volume per unit mass	cc/g or ft^3/lb_m
Specific gravity	G	ρ/ρ_{ref}	Dimensionless
Specials	°Bé	$G > 1: 145 - \dfrac{145}{G}$	Degrees Baumé
		$G < 1: \dfrac{145}{G} - 130$	
	°Tw	$\dfrac{G-1}{0.005}$	Degrees Twaddell
	°API	$\dfrac{141.5}{G_{60}} - 131.5$	Degrees API

The density is a physical property of any material (solid, liquid, or gas). It is the mass of some stated volume. In the metric system, it is the mass (g) of one cubic centimeter (cc) of the material. In the English system, it is the mass (lb_m) of one cubic foot (ft^3) of the material. Occasionally you will encounter the specific volume, which is the volume per unit mass. The specific volume is the reciprocal of the density.

The specific gravity is the ratio of the density of the material to the ᵥ of a reference material. For liquids, the reference material is usually, but always, water at $4\,°C$, which has a density of $1.0\,g/cc$. With this reference, the numerical value of the specific gravity is the same as the numerical value of the density in grams per cubic centimeters. But in the petroleum industry, the reference is more often water at $60\,°F$ $(15\,°C)$, which has a density of $0.9991\,g/cc$.

Certain industries have traditionally used other scales for specific gravity. In the petroleum industry, $°API$ is common (API is the American Petroleum Institute). In the formula, the gravity G_{60} is the density of the material at $60\,°F$ relative to the density of water at $60\,°F$.

Effect of Temperature

The density of a liquid is a function of temperature and composition. In most applications, the effect of pressure on liquid density is insignificant (liquids are incompressible). But there are exceptions, such as a steam boiler operating at $1500\,psig$. Water is slightly compressible at these pressures.

Values for densities and specific gravities must always be accompanied by a value for the temperature. In certain applications, there is a generally accepted or reference temperature at which densities and specific gravities are to be stated. As just noted, in the petroleum industry, it is generally $60\,°F$ $(15\,°C)$. For many materials, tables are available that state the density or specific gravity as a function of temperature. A commonly used correlation for density as a function of temperature is the Rackett equation:

$$\rho = A \times B^{-(1-T/T_c)^n}$$

where ρ = density (g/cc); T = temperature (K); T_c = critical temperature (K); A, B, and n = coefficients.

The values for the coefficients for many organic and inorganic compounds are readily available.[1]

Inventory Volume

When stating the volume of liquid in a tank, there are two bases:

Actual volume or process volume. This is the volume at process conditions— that is, at whatever the current process temperature happens to be. If you are concerned about exceeding the capacity of a vessel, this is the liquid volume of interest.

Inventory volume. This is the volume at the reference temperature. If you are reporting production rates, inventory amounts, etc., this is the liquid volume of interest.

This distinction applies only to volume; actual weight and inventory weight are the same.

Stratification

Ideally, the computation of mass from a level measurement is simple:

1. Compute liquid volume from level. The relationship depends on vessel geometry.
2. Multiply volume by density to obtain mass of liquid.

This is expressed as follows:

$$V = f(H)$$
$$M = \rho V$$

where H = level of liquid in vessel (cm or ft); V = volume of liquid in vessel (L or ft^3); M = mass of liquid in vessel (kg or lb_m); ρ = density of liquid in vessel (kg/L or lb_m/ft^3).

In the metric system, note that $1\,g/cc = 1\,g/mL = 1\,kg/L = 1000\,kg/m^3$.

Because density is a function of temperature, one must either

• Sense the density directly.
   OR
• Sense the temperature and infer density from temperature.

But in practice, a rather nasty issue arises: Are the contents of the vessel uniform? Very unlikely, especially for solids, but also for liquids in unagitated storage tanks. In the absence of agitation, liquids tend to stratify. Where do you measure the density? Where do you insert the temperature probe? The only way to avoid errors is to avoid stratification. For liquids, this means providing sufficient agitation in some form (such as a pump-around). For solids, this is very difficult to achieve.

Foam

Some materials have a tendency to foam. In a vessel, the foam layer is on top of the liquid surface. Agitation can lead to foaming, as can boiling, such as in evaporators. Establishing vacuum on a vessel can cause foaming (some low-boiling components vaporize at low pressures).

The quantity of liquid in the foam layer is very small and is not normally of interest. There are two issues:

• The foam is a potential problem for some level measurement technologies.

- Sometimes the height of the foam layer is of concern. For example, when pulling vacuum on a vessel, the foam must not enter the vacuum system.

Rag Layer

The *rag layer* pertains to liquid–liquid interfaces. Two liquids, such as oil and water, are said to be immiscible—that is, they will not form a mixture. With no agitation, they separate into two phases; the light (low-density) phase is on top and the heavy (high-density) phase is on the bottom.

A very common characteristic of liquid–liquid interfaces is a rag layer (solid–liquid interfaces behave in a similar fashion). The interface is fuzzy. There is a gradual transition from one phase to the other.

The rag layer presents problems for some level measurement technologies, but not all. For example, when inferring the interface level from pressure, the theoretical location of the "clean" interface can be calculated. For some applications, this is acceptable. For others, it is not. In some applications, the phases are allowed to separate so that the lower phase can be drained, leaving only the upper phase. If the objective is for the drained material to be lower phase material only (none of the upper phase), one must stop draining at the lower limit of the rag layer. If the objective is to leave only upper phase material in the vessel (none of the lower phase), one must continue to drain until the upper limit of the rag layer. The best approach is usually to drain the vessel through a coriolis flowmeter that also senses density and then terminate the drain based on the density.

Level-Sensing Technologies

The technologies for sensing level include the following:

Technology	Liquids	Solids
Float	Yes	
Displacer	Yes	
Hydrostatic head (pressure)	Yes	
Resistance tape	Yes	
Visual (dipstick; sight glass)	Yes	
Capacitance	Yes	Yes
Load cell	Yes	Yes
Nuclear	Yes	Yes
Noncontact radar	Yes	Yes
Guided wave radar	Yes	Yes
Ultrasonic	Yes	Yes
Vibrating element	Yes	Yes

All of the technologies are applicable to liquids but only certain ones can be used for solids. As compared to liquids, the behavior of solids is far more complex and can differ substantially from one solid to another. For liquids, the surface must be even (except for ripples from agitation), and there can be no empty cavities below the surface. This is not true for solids:

- When adding solids to a tank, a mound is created below the addition point. Equipping the solids feeder with a distributor gives a more uniform surface, but never as even as a liquid surface.
- When removing solids from the bottom of a vessel, solids drain preferentially from the middle of the vessel, creating an inverse cone effect.
- Some solids will *bridge*, creating an empty space, or *rat hole*, below the surface.

These problems pertain to all technologies that sense level at a point on the surface. Only load cells and nuclear devices are immune to these effects.

4.2. PRESSURE TRANSMITTER

About 75% of level measurements are based on either pressure or differential pressure. We shall examine pressure first, and then differential pressure.

Pressure measurements are appropriate only for open tanks such as shown in Figure 4.2. Sometimes these measurements are used for vented tanks and other closed tanks that should be at atmospheric pressure. But before taking such a measurement, one must ask this question: Is there any situation in which the pressure in the headspace is anything other than atmospheric? If the

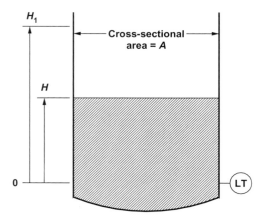

Figure 4.2. The physical location of the pressure transmitter corresponds to a value of zero for the level.

answer is yes, then the level indication based on the pressure measurement will be in error should this situation ever arise. The primary concern is that such a situation will occur during abnormal or upset conditions, when the process operators are very busy. Giving the operators a false level indication in such conditions is not a good idea. This can be avoided by using differential pressure.

In the subsequent examples, we will assume the following:

- All physical dimensions of the vessel are known.
- The specific gravity G (relative to water) of the liquid in the tank is known.
- The level will be measured in inches.
- The units for the pressure sensed by the pressure transmitter will be in in H_2O.

Simplest Configuration

In the vessel in Figure 4.2, a gauge pressure transmitter is installed near the bottom of the tank. The simplest configuration is when the pressure transmitter is physically installed at the point at which the level in the tank is considered to be zero.

The level measurement indicates levels up to H_1 in above the location of the pressure transmitter. The configuration parameters for the pressure transmitter must be as follows:

Value	Indicated Level	Pressure at Transmitter, in H_2O
Lower-range value	0	0
Upper-range value	H_1	GH_1
Span	H_1	GH_1

For such applications, the direct-mount configuration described in Chapter 3 should be considered. The pressure transmitter is supplied connected to a flange of the desired size (minimum size is about 2 in). A flush mount option designed specifically for level measurement is available, which avoids any dead space in the connection to the vessel.

Other Reference Points

Suppose production insists on indicating the level relative to the bottom of the vessel, as illustrated in Figure 4.3. Note that H_1 is now referenced from the bottom of the tank, not from the location of the pressure transmitter.

Because the pressure transmitter is located H_0 in above the bottom of the tank, the pressure transmitter cannot sense levels less than H_0. But if it could, what would the pressure be when the level is at the bottom of the

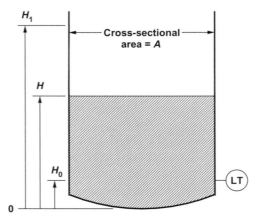

Figure 4.3. The physical location of the pressure transmitter does not correspond to a value of zero for the level.

tank? It would be $-GH_0$. To reference the level indication to the bottom of the tank, the configuration parameters for the pressure transmitter must be as follows:

Value	Indicated Level	Pressure at Transmitter, in H_2O
Lower-range value	0	$-GH_0$
Upper-range value	H_1	$G(H_1 - H_0)$
Span	H_1	GH_1
Minimum measurable level	H_0	0

Even though the measurement range extends to the bottom of the tank, the level indication can never be less than H_0.

Smart Transmitters

The configuration of a smart pressure transmitter to provide a measured value of the level is quite easy. The procedure is as follows:

Specify one point on the transmitter operating line. Usually this is at the lowest pressure that can be sensed by the pressure transmitter. This value would be sensed by the pressure transmitter under the following conditions:

- Vessel should be empty (or liquid level at or below the vessel connection for the pressure transmitter).

- The lines, if any, connecting the pressure transmitter to the vessel should be filled as in normal operations.

For the pressure sensed under these conditions, specify the value to be indicated for the level. If the physical location of the pressure transmitter corresponds to a level of zero, the sensed pressure will be zero and the indicated value for the level will be zero. But for the example in Figure 4.3, the sensed pressure will be zero, but the transmitter output should be H_0.

Specify the span. For the measured value, this will be the span in level units (H_1 for the examples in Figures 4.2 and 4.3). For the sensed pressure, multiply this height in inches by the liquid specific gravity to obtain the span for the sensed pressure in in H_2O (GH_1 for the examples in Figures 4.2 and 4.3).

This procedure can be applied when capillary seals, purged connections, etc. are present.

Although we shall do so in our examples, relating the pressure sensed for the empty vessel to the physical dimensions and other parameters of the transmitter installation is normally not necessary. But should the indicated level be unreasonable, one action item for troubleshooting is to verify that the pressure sensed for the empty vessel is consistent with the value computed from the physical dimensions of the installation.

Zero Suppression and Zero Elevation

The terms *zero suppression* and *zero elevation* date from the era when technicians directly specified the zero adjustment for conventional pressure transmitters. Because this is no longer the case with smart transmitters, these terms are encountered less frequently. In digital systems, the term *zero suppression* is frequently used in another context, specifically, for eliminating the nonsignificant zeroes (leading or trailing) from a numerical representation of a value.

When configuring a smart transmitter, any point on the operating line of the transmitter can be specified. This point consists of the value to be indicated for the level at a specified value of the sensed pressure. From this value and the span, the transmitter computes the transmitter's zero, which is value of the sensed pressure for which a value of zero is to be indicated for the level. If this sensed pressure is negative, the transmitter's zero is said to be suppressed. If this sensed pressure is positive, the transmitter's zero is said to be elevated.

For the installation in Figure 4.3, the transmitter is to output H_0 when the sensed pressure is zero. To output a value of zero for the level, the sensed pressure would have to be $-GH_0$. This is the required setting for the transmitter's zero, and because this value is negative, the transmitter's zero is said to be *suppressed*. As indicated previously, a negative sensed pressure is not

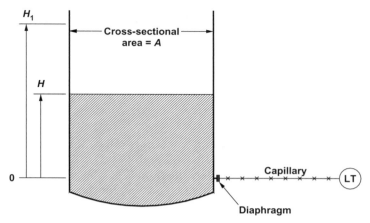

Figure 4.4. The physical location of the pressure transmitter with capillary seals corresponds to a value of zero for the level.

possible for the installation in Figure 4.3, so the minimum value indicated for the level will be H_0.

In some configurations, the transmitter is to indicate a level of zero when the sensed pressure is a positive value. For these cases, the transmitter's zero must be *elevated*.

Capillary Seal

Sometimes the pressure transmitter must be physically located some distance from the vessel, as illustrated in Figure 4.4. For example, the vessel temperature may be above the allowable limits for the pressure transmitter. A capillary seal system is one way to address this issue.

If the pressure transmitter is physically mounted at the same elevation as the connection to the vessel, the capillary seal system can be ignored. The capillary tube itself may be routed in whatever direction is required. But as long as the two ends are at the same elevation, its contribution to the head is zero. The measurement range for the pressure transmitter can be specified as if the pressure transmitter were physically located at the connection to the process vessel. The configuration parameters are the same as if the capillary seal system were not present:

Value	Indicated Level	Pressure at Transmitter, in H_2O
Lower-range value	0	0
Upper-range value	H_1	GH_1
Span	H_1	GH_1

Transmitter Below Tank Connection

Although locating the pressure transmitter at the same elevation as the tank connection gives the simplest result, someone will find a good reason not to do so. In Figure 4.5, the pressure transmitter is located at a distance H_C below the vessel connection. If the pressure transmitter is located above the vessel connection, the expressions that follow apply if the value of H_C is considered to be negative. The expressions must include the liquid head of the fluid in the capillary seal system. Let the specific gravity of the fluid in the capillary seal system be G_C. Because the liquid head is $G_C H_C$, the transmitter must be configured as follows:

Value	Indicated Level	Pressure at Transmitter, in H_2O
Lower-range value	0	$G_C H_C$
Upper-range value	H_1	$GH_1 + G_C H_C$
Span	H_1	GH_1

If one follows the procedure presented previously for configuring a smart transmitter, the sensed pressure (in in H_2O) with the vessel empty is the product of H_C and G_C. It is not necessary to measure H_C and obtain a value for G_C. But if problems are experienced with the indicated value of the level, one might want to verify that the sensed pressure with the vessel is empty is consistent with the values of H_C and G_C.

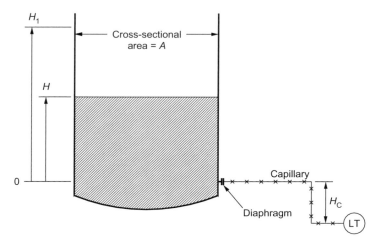

Figure 4.5. The physical location of the pressure transmitter with capillary seals does not correspond to a value of zero for the level.

Purged Connection

When the pressure transmitter must not come into contact with the process fluids, purging, as illustrated in Figure 4.6, should be considered. Purging may be with gas or liquid:

Gas. The specific gravity of gas is negligible compared to that of any liquid. The physical location of the pressure transmitter is irrelevant. The pressure transmitter is configured as if it were physically located at the vessel connection.

Liquid. The simplest arrangement is to physically locate the pressure transmitter so that it is at the same elevation as the vessel connection. The net head of the purge fluid is zero, and the pressure transmitter is configured as if it were physically located at the vessel connection. But if the pressure transmitter is at a different elevation, the head contributed by the purge fluid must be taken into consideration. The approach is exactly the same as for capillary seal systems.

Air Bubbler

For open tanks, the air bubbler illustrated in Figure 4.7 is a simple purged system. Air (or occasionally some other gas) is bled into the tank through a pipe or tube that extends to a known location below the liquid surface. The flow of air is judged by the rate at which the bubbles break the surface. At first impression, air bubblers seem to be a crude approach. But the flow of air keeps the tube clear of obstructions, so it works successfully in some quite dirty applications.

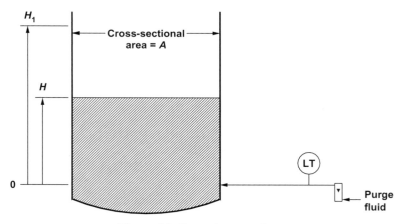

Figure 4.6. Purged connections.

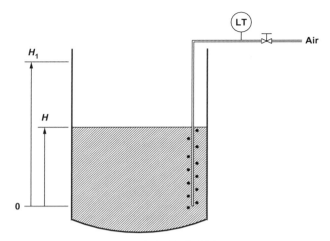

Figure 4.7. Air bubbler.

A pressure transmitter senses the pressure required to maintain the flow of air through the tube. This pressure is exactly the same as the pressure in the vessel at the tip of the tube. This pressure is the head due to the liquid in the tank.

The simplest configuration is when the tip of the tube is at the point in the vessel that corresponds to a level of zero. The configuration parameters for the pressure transmitter would be as follows:

Value	Indicated Level	Pressure at Transmitter, in H_2O
Lower-range value	0	0
Upper-range value	H_1	GH_1
Span	H_1	GH_1

Specific Gravity Errors

The specific gravity of the liquid in the tank is a function of temperature and composition. Temperature can be sensed and compensation applied, provided factors such as stratification can be resolved and a representative temperature obtained. Compensating for composition changes is far more difficult.

It is not always necessary to compensate for changes in specific gravity. The need depends on the ultimate measured variable:

Level. When inferring level from pressure, errors in the specific gravity lead to errors in the level.

Volume. The errors in the level lead to the same errors in liquid volume.

Mass. To obtain liquid mass, the liquid volume is multiplied by the density of the liquid. In many applications, this largely if not totally cancels the errors due to specific gravity. Consider a straight-walled vessel as illustrated in Figure 4.2. The errors in specific gravity would cancel for all liquid above the location of the pressure transmitter. However, they would not cancel for the liquid that is below that location. For vessels without straight walls, there is another contribution. The liquid volume is determined from the liquid level, either by formula or strapping table. Errors in the level lead to errors in the liquid volume. These are not entirely canceled when the liquid volume is multiplied by density to obtain liquid mass.

4.3. DIFFERENTIAL PRESSURE TRANSMITTER

To sense the level in a closed vessel requires a differential pressure measurement. The lower connection to the vessel establishes the lowest level that can be sensed; the upper connection establishes the highest level that can be sensed.

One component of the differential pressure is the liquid head that varies with the level in the vessel. However, there is another potential component that depends on the contents of the connection piping (or "legs") to the differential pressure transmitter. In the configuration illustrated in Figure 4.8, the differential pressure transmitter is physically located at the lower connection to the vessel. There are two possibilities for the connection piping, or leg, to the upper connection:

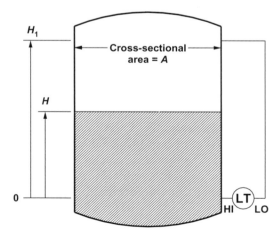

Figure 4.8. The physical location of the differential pressure transmitter corresponds to a value of zero for the level.

Dry leg. The connection piping to the upper connection contains only gas. There would be no contribution to the differential pressure. However, the installation must be such that no liquid can accumulate in this leg.

Wet leg. The connection piping to the upper connection is completely filled with a liquid. The resulting liquid head is a component to the differential pressure. This can be taken into account, provided the installation is such that it is known with certainty what this liquid is.

Smart Transmitters

With smart transmitters, the procedure for configuring the differential pressure transmitter is basically the same as that presented previously for configuring the pressure transmitter. This procedure is as follows:

Specify one point on the transmitter operating line. The differential pressure is sensed by the transmitter under the following conditions:
- Vessel should be empty (or liquid level at or below the lower vessel connection for the differential pressure transmitter).
- The legs are filled as in normal operations.

For the pressure sensed under these conditions, specify the value to be indicated for the level.

Specify the span. For the measured value, this will be the span in level units. Multiply this height in inches by the liquid specific gravity to obtain the span for the sensed differential pressure in in H_2O.

This procedure can be applied for wet legs or dry legs.

Dry Leg

Assume that the piping to the upper connection is free of liquids and that the following are known:

- The distance (in) between the vessel connections, or H_1 as shown in Figure 4.8.
- The specific gravity G (relative to water) of the liquid in the tank. In regard to errors in the value of the specific gravity, the observations previously made for pressure transmitters also apply to differential pressure transmitters.

The level measurement can indicate levels from the location of the lower connection up to H_1 inches above that location. For dry legs, the lower vessel connection is connected to the HI side of the differential pressure transmitter, and the upper connection is connected to the LO side. The configuration parameters for the differential pressure transmitter must be as follows:

Value	Indicated Level	Pressure at Transmitter, in H_2O
Lower-range value	0	0
Upper-range value	H_1	GH_1
Span	H_1	GH_1

Condensate Pot   Suppose the vessel contains a liquid at a temperature well below its boiling point. If the pressure transmitter is physically located at the lower connection, would the piping to the upper connection remain dry? Unfortunately, no. Condensates tend to form in such connection piping, but at a rather slow rate. Unless some provision is made, they will accumulate over time, and the resulting liquid head leads to an error in the level measurement.

When the condensation rate is quite slow, a condensate pot is usually a simple solution to this problem. As shown in Figure 4.9, the condensate pot is merely a small vessel attached to the piping to the upper connection. This piping must be sloped such that any condensate that forms drains into the condensate pot, and the liquid must be periodically drained from the pot.

Condensation rates are a function of temperature, so one would expect more in cold weather. One way to slow the condensation rate is to heat trace the piping to the upper connection. In cold climates, heat tracing of the condensate pot and the dry leg may be necessary to prevent freezing.

Purge Gas   Another way to ensure that the connection piping in the dry leg is free of liquid is to purge with a suitable gas (Fig. 4.10). Such installations are attractive when the condensate is corrosive, tends to solidify and plug the lines, and so on.

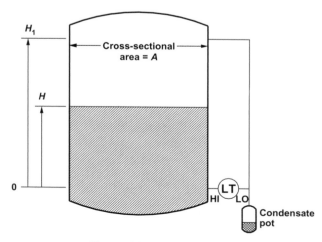

Figure 4.9. Condensate pot.

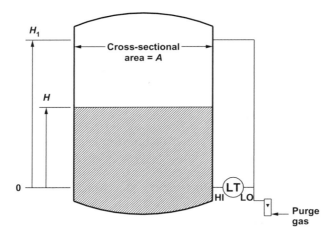

Figure 4.10. Purged connection for dry leg.

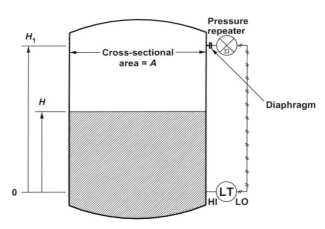

Figure 4.11. Pressure repeater.

The main down side of purging is that it must be maintained at all times. Especially in corrosive applications, even temporary loss of purge can have serious consequences.

Pressure Repeater As illustrated in Figure 4.11, a pressure repeater is installed at the upper connection to the vessel. The pressure repeater senses the pressure in the vessel via a diaphragm and outputs a pressure that is the same as the pressure in the vessel. The dry leg now contains instrument air.

Being a pneumatic component, the pressure repeater requires a supply of instrument air at the appropriate pressure. The output pressure from the pressure repeater cannot exceed the pressure of the supply air. One must determine the maximum pressure expected in the vessel and provide a supply air that is greater than this pressure. Although pneumatic instrumentation and controls are rarely installed today, pressure repeaters are still commercially available.

Wet Leg

Assume that the piping to the upper connection is completely filled with a liquid and that the following are known:

- The distance (in) between the vessel connections, or H_1.
- The specific gravity G (relative to water) of the liquid in the tank.
- The specific gravity G_L of the liquid in the wet leg.

With no liquid in the tank and the HI connected to the lower vessel connection, the differential pressure will be $-G_L H_1$. As the tank fills with liquid, the differential pressure increases (becomes less negative). If $G < G_L$, the differential pressure will never be positive. The differential pressure transmitter must be configured as follows:

Value	Indicated Level	Pressure at Transmitter, in H_2O
Lower-range value	0	$-G_L H_1$
Upper-range value	H_1	$H_1 (G - G_L)$
Span	H_1	GH_1

Seal Pot   In some applications (such as steam boilers), the vessel contains a liquid at or near its boiling point. Therefore, considerable condensate will form in the piping to the upper vessel connection. The condensation rate is so rapid that it is not possible to achieve a dry leg by draining the condensate into a condensate pot.

The other alternative is to make sure that the piping to the upper connection is completely full of condensate. This is the purpose of the seal pot illustrated in Figure 4.12. The seal pot is installed at the upper connection to the vessel and provides a reservoir of liquid to ensure that the piping is completely full of condensate. Condensate forms in the seal pot and largely drains back into the vessel. The expectation is that heat is continuously being lost from the seal pot, and thus condensate is continuously forming in the seal pot. If required, heat tracing should be applied carefully; excessive heating of the seal pot interferes with its purpose.

Seal pots are commonly installed in steam boilers and similar applications in which the condensate will be a single component. When the condensate is a mixture, questions arise as to the composition of the liquid in the wet leg.

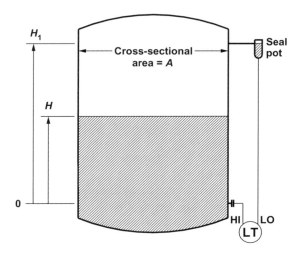

Figure 4.12. Seal pot for wet leg.

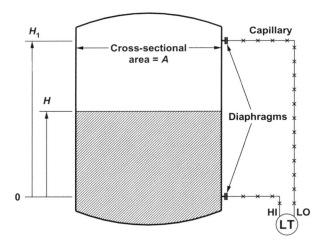

Figure 4.13. Capillary seal systems.

Capillary Seal Systems The use of capillary seal systems on both sides of the differential pressure transmitter permits the transmitter to be located anywhere. The differential pressure transmitter in Figure 4.13 is physically located at some distance below the lower vessel connection. It could even be somewhat above the lower vessel connection. It is not necessary to know how far above or below.

No matter where the differential pressure transmitter is located, the net contribution of the liquid head for the capillary seal systems is $G_L H_1$. This is the differential pressure sensed by the transmitter when the vessel is drained of liquid.

4.4. CAPACITANCE AND RADIO FREQUENCY

A capacitor consists of two conducting surfaces separated by a dielectric material—that is, a material that can store and release electrons but is not a conductor of electrons. The two relevant material properties are

Dielectric constant K. The permittivity of a material relative to the permittivity of a vacuum. *Permittivity* is a measure of the ability of a material to store an electrical charge. When subjected to a potential gradient, the material stores electrons. When the potential gradient is removed, the material releases electrons.

Conductivity. A measure of the ease with which electrons flow through a material. It is the reciprocal of resistivity. Resistivity is measured in ohm centimeters ($\Omega \cdot$cm). The units of conductivity are siemens per meter (formerly mho/m), where siemen = mho = ohm^{-1}.

The dielectric constant applies to solids, liquids, and gases. However, the dielectric constant of most gases is very close to 1.0. Following are the dielectric constants of selected materials:

Material	Dielectric Constant
Air	1.0
Benzene	2.3 (68 °F)
Ethanol	24.3 (77 °F)
Gasoline	2.0 (72 °F)
Glass (silica)	3.8
Polyethylene pellets	1.5
Sulfuric acid (100%)	84.0 (68 °F)
Water	88.0 (32 °F)
	80.4 (68 °F)
	55.3 (212 °F)

The dielectric constant is a function of temperature, and generally decreases with temperature. For mixtures (such as acids), it is a function of composition. For many solids, it is strongly influenced by moisture content.

Level From Capacitance

The capacitance C of a capacitor is given by the following expression:

$$C = \frac{\varepsilon_0 K A}{D}$$

where C = capacitance of the capacitor; A = area of the plates; D = distance between the plates; ε_0 = permittivity of free space (8.85×10^{-12} Fs/m).

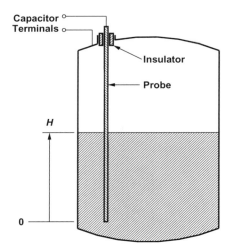

Figure 4.14. Using capacitance to measure level, nonconducting liquid.

To measure level, we generally construct the capacitor by inserting a conducting probe into a metallic tank, as illustrated in Figure 4.14. If the probe is insulated from the tank, the probe is one plate of the capacitor and the tank is the other.

Capacitance can be computed for any geometry, but we can avoid that by measuring the capacitance when the tank is empty; let this be C_0. We then fill the tank with a material with dielectric constant K. We have displaced a material (air) with a dielectric constant of 1.0 with a material whose dielectric constant is K. Therefore, the capacitance should be KC_0. Assuming the tank walls are straight, the capacitance varies linearly from C_0 to KC_0 as we fill the tank.

Conducting and Nonconducting

A fluid with a conductivity of 10 microsiemens/cm or greater is considered to be conducting. It is interesting to note that nonconductive fluids generally have low dielectric constants, and conducting fluids have high dielectric constants. It turns out that materials with a dielectric constant of 10 or greater are usually conducting (have a conductivity of 10 microsiemens/cm or greater).

To have a capacitor, the media between the plates must be nonconducting. When the fluid is conducting, the probe must be coated with an insulating material, as illustrated in Figure 4.15. The tank wall is grounded; the conducting fluid within the tank will also be at ground potential. The dielectric for the capacitor is now the probe insulation. The dielectric constant of the probe insulation determines the capacitance; the dielectric constant of the fluid has no effect on the capacitance. In such applications, coating the probe with a material that has a high dielectric constant provides a greater sensitivity.

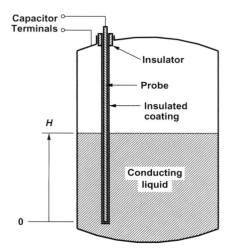

Figure 4.15. Using capacitance to measure level, conducting liquid.

The value of the capacitance is determined by the dielectric constant of the probe insulation and the amount of the insulation that is covered by the conducting liquid. This capacitance varies linearly with level.

Impedance and Admittance

In the DC world, the ratio of voltage to current is resistance. A capacitor offers infinite resistance. Imposing a voltage will provide a current flow while the capacitor is charging, but at steady state, the current flow will be zero.

The capacitance must be sensed using an AC voltage (usually referred to as a *high-frequency oscillator*). The relationships between voltage V, current I, frequency f, capacitance C, resistance R, impedance Z, and admittance A are as follows:

Result	Resistor	Capacitor
Current	$I_R = \dfrac{V}{R}$	$I_C = C\dfrac{dV}{dt}$
DC: $V = 1$	$I_R = \dfrac{1}{R}$	$I_C = 0$
AC: $V = \sin 2\pi ft$	$I_R = \dfrac{\sin 2\pi ft}{R}$	$I_C = 2\pi fC \cos 2\pi ft$
Impedance	$Z_R = R$	$Z_C = \dfrac{1}{2\pi fC}$
Admittance	$A_R = \dfrac{1}{R}$	$A_C = 2\pi fC$

When an AC voltage of frequency f (cycles/sec, or Hz) is imposed on a capacitor, a current I_C flows as the capacitor is alternately charged and discharged. This current flow increases with capacitance C and frequency f. In the AC world, the ratio of voltage to current is referred to as the impedance Z (admittance is merely $1/Z$). For a resistor, the impedance is the resistance. For a capacitor, the impedance decreases with capacitance C and with frequency f.

In real-world level measurement applications, the dielectric material will have some conductivity. Therefore, the current sensor will detect the sum of the resistive current I_R and the capacitive current I_C. The resistive current I_R is independent of frequency f, but the capacitive current I_C increases with frequency f. When high frequencies are used, the resistive current I_R is negligible compared to the capacitive current I_C—that is, $I_C \gg I_R$. In level measurement applications, these frequencies are in the RF (radio frequency) range (for example, the Siemens SITRAN LC 500 operates at 420 kHz).

Capacitive and Resistive Impedance

Impedance is actually a vector quantity—that is, it has a magnitude Z and phase angle ϕ. Figure 4.16 illustrates the results of applying 1.0 V AC to a resistor and a capacitor. For the resistor, the current is in phase with the voltage (that is, the peaks occur at the same time). But for the capacitor, the peaks do not align. When the voltage is at its peak, the current flow is zero; the current flow is at its peak when the voltage is zero. The peak in the current occurs one quarter cycle later than the peak in the voltage. Because a full cycle is 360°, the current is 90° out of phase with the voltage.

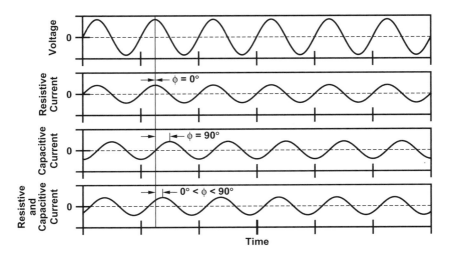

Figure 4.16. Resistive and capacitive impedance.

When the impedance is the result of both resistance and capacitance, the phase angle of the current will be between $0°$ and $90°$ relative to the voltage. By measuring the magnitude and the phase angle of the current, the impedance can be resolved into resistive impedance Z_R and capacitive impedance Z_C. For level measurement, it is the capacitive impedance Z_C that is of interest.

RF Capacitance Level Measurement

Some suggest that capacitance level measurement devices have been replaced with RF level measurement devices. Actually, capacitance level measurement devices evolved into RF level measurement devices. Some use the term *RF impedance*; others use the term *RF admittance*. All refer to the same basic technology (remember that admittance is the reciprocal of impedance). For continuous level measurement, there are two components:

Capacitance probe. This must be inserted into the process vessel. Therefore, capacitance level measurement technology is intrusive. However, there are no moving parts, and in most cases, the probe is inserted through a single opening on the top of the vessel. Short probes will be rigid, but long probes can be flexible (such as a chain or cable with a weight on the end).

Transmitter. This unit imposes an AC voltage at RF frequencies, senses the current, computes the impedance, and outputs a current loop signal or a digital value that indicates the level in the vessel. The introduction of microprocessor technology into the transmitter has improved the signal processing (such as resolving total impedance into resistive impedance and capacitive impedance) and simplified calibration. The transmitter is usually directly mounted, but the transmitter can be remotely mounted if necessary.

Buildups

Buildups on the probe have presented problems for capacitance level measurement devices since the very beginning. Buildups on the probe offer both capacitance and resistance. Consequently, a device that measures only total impedance will respond to both effects, the result being a level indication that is higher than the actual level. Going from the traditional capacitance technology to RF technology offers some improvement, but only because the resistive current is negligible compared to the capacitive current at RF frequencies. But with a microprocessor in the transmitter, additional steps are possible, and most manufacturers have incorporated features designed to address this problem.

One approach is based on the observation that conductive materials have high dielectric constants and nonconductive materials have low dielectric constants. As the buildups on the probe increase, the resistive current and the capacitive current both increase. The resistive current increase is due to the conductance of the buildups on the probe. Therefore, the capacitance of the buildups on the probe should increase along with the conductance. The resistive current increase from the buildups can be translated into a corresponding capacitive current increase from the buildups. Subtracting this capacitive current increase from the capacitive current that is sensed gives the net capacitive current for the contents of the vessel.

Variations in the Dielectric Constant

Most commercial products provide the capability to compensate for the effect of temperature on the dielectric constant. But compensating for composition changes is more ambitious. For some solids, the moisture content has a significant effect on the dielectric constant.

If one knew the composition, one could compensate for the effect of composition on the dielectric constant. However, sensing the composition is usually an ambitious undertaking. In most applications, it would be easier to sense the dielectric constant. The technology used to infer level from capacitance can also be used to infer the dielectric constant from capacitance.

To do so, a capacitor of known geometry must be inserted into the vessel at a location that is fully submerged at all times. Figure 4.17 illustrates the possibilities:

- Use the lower portion of the probe to sense the capacitance of the material in the vessel. For example, Drexelbrook's True Level III uses the lower 5.5 in of the probe for this purpose.

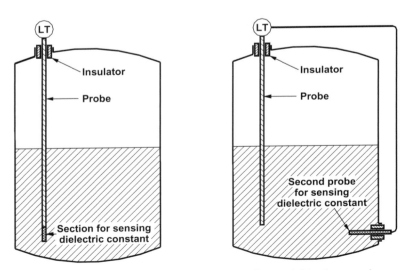

Figure 4.17. Sensing dielectric constant of material in the vessel.

- Insert a second probe into the vessel. The transmitter then senses the capacitance. The geometry of this capacitor is known, so the dielectric constant of the material can be inferred from its capacitance.

In both cases, the sensed capacitance gives the dielectric constant of the material at the location of this probe, which could be different from the remaining material in the vessel. Stratification can exist in nonagitated liquids, but solids pose the greatest concern.

Calibration

Calibration of conventional capacitance level transmitters generally required two steps (and in this order):

1. Drain the vessel to below the tip of the probe. Instruct the transmitter to sense the capacitance, and use this as the lower-range value for the capacitance.
2. Fill the vessel to a known level. Specify this level to the transmitter, and then instruct the transmitter to sense the capacitance. In the early versions of some products, this level had to be at the upper-range value of the level measurement.

For simple vessel geometries, linear interpolation is then applied to convert the sensed capacitance to vessel level.

Smart transmitters have relaxed this procedure considerably. The two-point calibration can be undertaken with any two points (capacitance sensed by transmitter at a known level). And they can be in any order. And instead of a second point, the sensitivity of capacitance with respect to level can be directly specified to the transmitter. And then what at least sounds like the ultimate, those that sense the dielectric constant can even be self-calibrating.

Advantages and Disadvantages

The major advantages of capacitance probes are as follows:

- Applicable to solids.
- No moving parts.
- High-temperature, high-pressure probes are available.

The disadvantages are as follows:

- Intrusive, but through a single opening.
- Dielectric constant must be known or sensed.
- Buildups are a potential problem.

When capacitance probes first appeared, their being applicable to solids was a major advantage. At that time, the only alternatives were nuclear level measurement devices and load cells. But today, ultrasonic and radar are also options for measuring the level of solids in a bin.

Although intrusive, there are no moving parts. Only the probe is exposed to the process materials. Uninsulated probes can be manufactured from any metal with the necessary corrosion resistant characteristics. Probes can be insulated with either a suitable polymer or possibly a ceramic material for high temperatures. The probe does not have to be rigid; it can be a chain or cable with a weight on the end. Special probe designs address high-temperature and high-pressure applications. In most applications, the transmitter is physically mounted on the upper end of the probe; however, it can be remotely mounted if necessary.

Normally the tank wall provides one of the plates for the capacitor. However, nonmetallic tanks will not do so. One approach is to insert a second rod that is maintained (with insulated spacers) at a fixed distance from the capacitance probe. Some metallic tanks are lined. A conducting liner has no effect, but an insulating liner (such as rubber) will contribute to the capacitance sensed by the probe.

Buildups continue to be a cause for concern. Manufacturers quickly became aware of this problem. RF technology offered some improvement. Microprocessors permit more logic to be incorporated to address this issue. However, the problem seems to persist.

4.5. ULTRASONIC

As illustrated in Figure 4.18, an ultrasonic level sensor transmits a burst of sound in the direction of the surface and then determines the time for the reflection to arrive at the receiver. This *time of flight* is given by the following expression:

$$\text{Time-of-flight} = \frac{2D}{v}$$

where D = distance from the transmitter to the surface; v = sonic velocity in the medium.

The ultrasonic level transmitter can be installed either

Above the surface. The sound is transmitted through the gas in the direction of the surface. The sonic velocity in the gas must be known. This approach can be applied to measure both liquid and solid levels. The discussion herein will be directed to this type of installation.

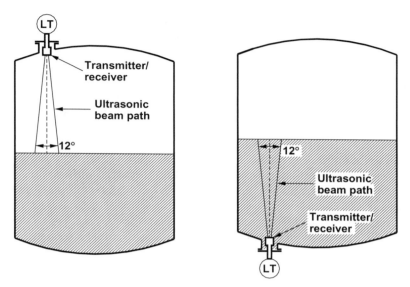

Figure 4.18. Ultrasonic level transmitter.

Below the surface. In liquid level measurement, the sound is transmitted through the liquid in the direction of the surface. The sonic velocity in the liquid must be known. This configuration is analogous to the sonar technology used in ships. But for obvious reasons, this version is not very popular in industrial applications. We shall say no more about it.

Sonic Velocity

The expression for the sonic velocity in a gas is as follows:

$$v = \left[\frac{\gamma P}{\rho} \right]^{1/2} = \left[\frac{\gamma R T}{M} \right]^{1/2} \text{ for ideal gas}$$

where v = sonic velocity (ft/sec); P = absolute pressure (atm); T = absolute temperature (°R); M = molecular weight of gas (lb_m/lb-mol); R = gas law constant (0.7302 atm-ft^3/lb-mol-°R); γ = ratio of specific heats (constant pressure/constant volume); ρ = fluid density (lb_m/ft^3).

For an ideal gas, observe that the sonic velocity

- Increases with the square root of the absolute temperature.
- Is independent of pressure.
- Depends on gas molecular weight, and therefore the gas composition.

Although pressure does not affect the sonic velocity, sound is not transmitted through a vacuum. The low pressure limit on ultrasonic level measurement devices is typically −10 psig.

In air at 32 °F (0 °C), the speed of sound is 1087 ft/sec (331 m/sec); at 68 °F (20 °C), the speed of sound is 1129 ft/sec (344 m/sec). Therefore, it is essential that ultrasonic level measurement devices compensate for the effect of temperature on the sonic velocity, and all commercial products do so. In applications in which the gas composition varies, there is another source of error. This is a more serious problem because the gas composition is unlikely to be known. Probably the best solution is for the instrument to automatically determine the sonic velocity by timing the reflection from a target at a known distance into the vessel.

Blocking Zone

The blocking zone is also called the *near zone* or *dead zone*. Modern ultrasonic level sensors use a piezoelectric crystal to generate the burst of sound through its natural frequency of vibration. When power is removed from the crystal, the vibrations decay rapidly but not instantaneously. Consequently, the receiver cannot be activated until these have decayed sufficiently.

The consequence for level measurement is that there is a minimum required distance from the transmitter to the surface. In most commercial products, this is about 1 ft. The near zone is the distance between the transmitter and the maximum surface level that can be sensed.

In most installations, the transmitter is inserted through a nozzle so that it is positioned just below the top of the vessel. But if levels must be sensed within the resulting near zone, the transmitter cannot be installed in this manner. The alternative is to install the transmitter in a pipe well, as shown in Figure 4.19. The required diameter for the pipe well depends on its length. The manufacturer of the ultrasonic level transmitter provides the necessary data.

Installation

Selecting the location for mounting the sensor on the vessel is a crucial part of the installation. The requirements are

- Beam must be perpendicular to surface.
- Avoid vessel internals, such as fill lines.
- Not too close to wall.
- Away from sources of noise.

The objective is for the sensor to receive a clear echo from the surface. This is certainly not assured:

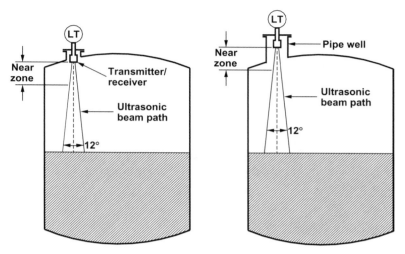

Figure 4.19. Near zone.

Multiple echos. The receiver is very likely to receive multiple echoes, one of which comes from the surface. The problem is to select the proper one. Some echoes will come from vessel internals, such as baffles, structural members, level switches, and agitator blades. Other echoes will be from the surface to the vessel wall to the sensor.

Echo loss. This occurs when the receiver is unable to detect an echo that is believed to be from the surface. This is a fault condition and is normally indicated as such. The output can be specified to fail upscale or downscale on this condition. There are several possibilities that can lead to this condition. In solids bins, the angle of the surface is not constant, so loss of echo may mean that the beam is no longer perpendicular to the surface. Some surface conditions cause the sound to be dispersed or absorbed instead of reflected. Dust on the surface in a solids bin can disperse the sound; agitation can cause a liquid surface to do likewise. Foam on a liquid surface can absorb the sound. Dust or liquid particles in the gas can also disperse the transmitted and/or reflected sound.

Advantages and Disadvantages

The major advantages of ultrasonic level transmitters are as follows:

- Measurement ranges up to 200 ft and more.
- Noncontact.
- Essentially nonintrusive from a single top entry.
- No moving parts.

The disadvantages are as follows:

- Echo loss can occur.
- False level indication due to extraneous echoes.
- Typical sensor temperature limit is −40° to 80 °C.
- Sonic velocity in gas must be known.

Ultrasonic level sensors use frequencies in the range of about 10 kHz up to about 60 kHz. Low frequencies (long wavelengths) are used to achieve measurement ranges of 200 ft or more; high frequencies (short wavelengths) are used for small tanks to achieve better resolution (as low as 1 mm). High-frequency units have even been used to sense the level in 55-gal drums. Some units permit the frequency to be adjusted; others use a selectable frequency, depending on the measurement range. Accuracy tends to be 0.25% to 0.5% of the upper-range value, but with resolutions of 1 to 6 mm.

The sonic transmitter/receiver is exposed to the temperatures at the top of the vessel. The typical temperature limits of −40° to 80 °C are too restrictive for some applications. The level transmitter module containing the microprocessor can be remotely mounted, but the transmitter/receiver cannot.

The main concern in most applications is maintaining a steady and reliable echo. The use of microprocessors in the transmitters enabled the manufacturers to provide logic designed to more reliably discern the echo from the surface from all the others. Some manufacturers also use the echo from an empty tank (tank mapping, tank profiling, etc.) to help the microprocessor in rejecting echoes from vessel internals. Most units provide a number of adjustable parameters. These permit the logic to be "tuned" to an application; however, remember that someone has to do the tuning, and that might be you.

4.6. NONCONTACT RADAR

Much like other radars, radar level sensors determine the distance to a target from the reflections or echoes. They are often referred to as *microwave level sensors* because they operate at frequencies typically between 5 and 25 GHz (gigaHertz or 1000 MHz), which are in the microwave region of the spectrum.

In many respects, the installation of radar level sensors is similar to that of ultrasonic level sensors. As illustrated in Figure 4.20, a transmitter/receiver (antenna) is mounted in the top of the vessel. The microwave transmitter directs a beam toward the surface of interest, and the receiver detects the reflection (or unfortunately, the reflections). Depending on the application, the antenna can be

Exposed. The antenna is subject to corrosion, buildups, and other realities of the industrial world. Technically, these are intrusive, but not excessively so.

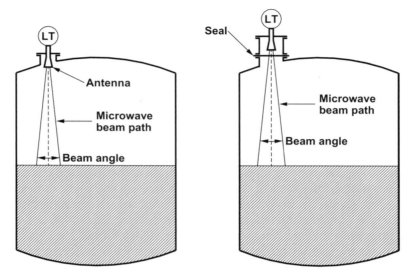

Figure 4.20. Level measurement by noncontact radar.

Sealed or isolated. Microwaves go through most plastics (and other materials with low dielectric constants). The antenna can be recessed and isolated from the process by a plastic seal. The antenna remains clean, but buildups on the process side of the plastic seal are sometimes an issue. These installations are nonintrusive, but at the expense of a larger vessel opening.

Time of Flight

Time-of-flight radar level sensors approach the problem in a manner similar to the ultrasonic level sensors. A microwave burst is directed to the surface, and the time required for the reflection to arrive at the receiver is determined. The relevant equation is the following:

$$\text{Time of flight} = \frac{2D}{c}$$

where D = distance from the transmitter to the surface; c = speed of light (186,000 m/sec).

For the distances encountered in level measurement, the times are extremely short. If the surface is 10 ft from the antenna, the time of flight would be 20.4 nsec (light travels approximately 1000 ft in 1 µsec or 1 ft in 1 nsec). Precisely measuring such times is quite a challenge. However, radar level sensors based on time of flight can provide the accuracy and resolution required for most industrial-level measurement applications.

One of the limitations of ultrasonic-level technology is that the speed of sound depends on the properties of the gas through which the sound travels. The speed of light is a universal constant. Its intensity can be attenuated on passing through a material, but its speed is not affected by media properties such as temperature, pressure, and composition.

Frequency Modulated Continuous Wave (FMCW)

The microwave signal frequency is varied in the fashion shown in Figure 4.21. Nominally, the transmitter operates at 10 GHz. However, the frequency of the transmitted signal varies in a linear sweep from 10 GHz up to 11 GHz.

As illustrated in Figure 4.21, the reflection exhibits the same sweep, but at a slightly later time. At any instant of time, the frequency of the received signal will be lower than the frequency of the transmitted signal. For a fixed sweep interval and a linear frequency sweep, the frequency difference is related to the time of flight. Instead of sensing the time of flight directly, the difference in frequency between the transmitted signal and the received signal is sensed. Very efficient numerical routines known as the fast Fourier transform (FFT) can be applied to determine precisely the difference in the frequencies. The frequency difference can be converted to the time of flight, from which the distance to the surface can be computed.

Radar level transmitters that rely on the FMCW technology have superior performance compared to those that directly sense time of flight. For tank gauging, custody transfers, and similar applications, the accuracy of the radar level transmitter is crucial, which suggests FMCW technology. But the counter-argument begins by determining the level of accuracy required for your application. The time-of-flight technology has improved to the point that it is usually sufficiently accurate for process applications.

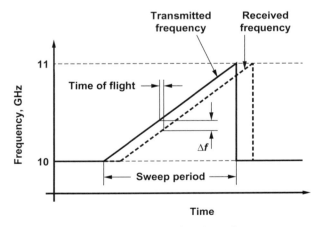

Figure 4.21. Frequency modulated continuous wave.

However, products based on the FMCW technology operate at higher frequencies (10 GHz for the example in Figure 4.21; some operate at 26 GHz). For a given measuring distance, this translates into a smaller antenna and a smaller beam angle. The influence of foam and dust on the reflection from the surface also depends on the frequency. The higher frequency will be advantageous in some applications but not necessarily all applications.

Multiple Reflections

The surface of interest is not the only source of reflections. Vessel internals such as baffles and structural members are sometimes excellent reflectors. Any structural members at right angles become *corner reflectors*. As illustrated in Figure 4.22, no matter what angle the beam impinges on a corner reflector, it is reflected directly back to its source.

If the incorrect reflection is selected, the radar level measurement system will indicate an incorrect value for the level. To avoid this, most suppliers of radar level measurement devices include tools specifically designed to reliably identify the reflection of interest from the other reflections. This entails quite sophisticated signal processing, which is now preformed by microprocessors. To understand what is causing the radar measurement device to have difficulty identifying the surface reflection, one needs to examine the echo pattern or profile for the empty tank. One then associates echoes with specific items (baffles, agitators, coils, etc.) within the tank. In some products, the microprocessor relies on this profile to reject echoes from the objects within the vessel.

When permissible, installing the radar level sensor in a stilling well reduces the stray reflections. Small vessels seem to present more problems than large ones. Glass-lined reactors are often very challenging.

No Reflection

When there is no reflection, the microprocessor is unable to detect the reflection from the surface. The culprits could be the following.

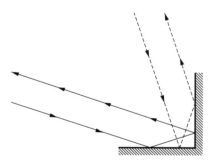

Figure 4.22. Corner reflector.

- A material with a high dielectric constant reflects most of the radar beam; a material with a low dielectric constant reflects very little. Regarding the dielectric constant required for a successful application, the line in the sand is a bit fuzzy, but it is around 1.8 or 1.9.
- Mist and dust particles in the atmosphere disperse the radar beam.
- Some foams absorb the radar beam. This is a difficult one. Some foams are transparent to the radar beam and do not interfere with the level measurement. Other foams absorb the radar beam.
- Surface turbulence from agitation scatters the radar beam.

Most microwave level measurements operate in the range of 5 to 25 GHz. Higher frequencies give a smaller beam angle for a given antenna or, conversely, a given beam angle can be attained with a smaller antenna. However, mist and dust tend to disperse the higher frequencies more than lower frequencies. Foam is more likely to absorb at the higher frequencies.

Antenna

The larger the antenna, the greater its sensitivity and the narrower the beam angle. The required size of the vessel nozzle is directly related to the size of the antenna. Except when a seal is required to isolate the antenna from the process, the antenna extends slightly into the vessel. The installation of the antenna must be with strict adherence to the manufacturer's recommendation. Four antenna types are illustrated in Figure 4.23:

- *Horn or cone.* This requires a 4-in or larger flanged vessel connection and can also be separated from the process via a plastic seal. Beam angle

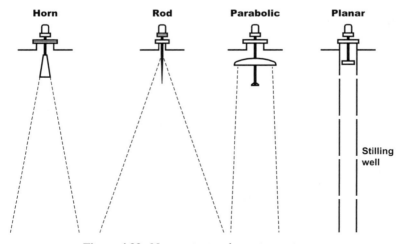

Figure 4.23. Noncontact radar antenna types.

depends on the size of the antenna (30° for small horns; 15° for large horns). The measuring range depends on size of horn but is typically at least 20 m (67 ft). The larger the horn, the better.

- *Rod or stick.* This requires a 2-in screwed or flanged vessel connection. Having the largest beam angle (30°) and lowest antenna sensitivity, it is limited to shorter distances with some loss in accuracy. The maximum measuring range can be as short as 10 m (33 ft).
- *Parabolic.* This requires the largest tank opening (basically a manhole, not a nozzle). However, it provides the smallest beam angle (6°) and has the highest antenna sensitivity, both of which are desirable for sensing levels at long distances. The measuring range can be 40 m (133 ft) or more.
- *Planar.* This is intended for still well installation.

Advantages and Disadvantages

The major advantages of noncontact radar for level measurement are as follows:

- High accuracy.
- Measurement ranges of 200 ft or more can be attained.
- Noncontact, no moving parts, not overly intrusive.
- Immune to vapor–space properties.

The disadvantages are as follows:

- Expensive.
- Does not work on low dielectric materials.
- Multiple reflections.
- Two-wire versions (low power) sacrifice performance.

The major advantage of radar level transmitters is their accuracy. Sensing the level with a resolution of 1 mm is indeed impressive.

Unfortunately, there are some materials whose dielectric constant is too low to adequately reflect microwaves. For example, propane has a dielectric constant of 1.6, which is too low. This is potentially a problem for any oil-like material. Radar will definitely work on water and anything that is aqueous-based or water-like.

The temperature and pressure limits rarely prohibit the installation of radar measurement devices. Most products tolerate process temperatures of up to 200 °C; some go much higher. Although standard products have lower ratings, those designed for high-pressure applications can tolerate 1000 psig or more.

Although prices continue to decline, radar level measurement devices are still costly. Two-wire intrinsically safe models are available, but the four-wire models deliver better performance and are capable of longer measuring distances.

4.7. GUIDED WAVE RADAR

Guided wave radar devices for industrial level measurement are based on a technology known as time domain reflectometry (TDR): Although this technology has been around for some time, applications were slow to develop due to the cost of the equipment and the complexity of its use. Advancements in semiconductor technology have reduced the costs, and microprocessors have simplified its use. It has now become the standard tool for locating faults in conducting cables. This technology is also being used for measurement of liquid and solid levels.

A pulse of energy is transmitted down a conductor inserted into the vessel, as illustrated in Figure 4.24. Wherever a change in impedance is encountered, part of this energy will be reflected. The greater the change in impedance, the greater the reflected energy. A change in impedance is associated with a fault in the conductor. A change in impedance occurs at a gas–liquid or a gas–solid interface. Both cause part of the energy to be reflected. To determine the distance from the transmitter to the fault or material surface, the time between transmitting the burst of energy and receiving the reflection must be sensed.

Because the energy travels at the speed of light, extremely short times must be sensed in level measurement applications. An additional problem is that there will be multiple reflections.

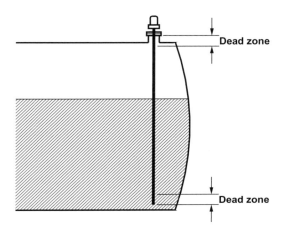

Figure 4.24. Level measurement by guided wave radar.

Multiple Reflections

Any change in impedance will reflect part of the energy. The reflection of interest is from the surface. Another reflection will be from the end of the conductor. In an oil–water separator, there are three reflections:

- From the air–oil surface.
- From the oil–water interface.
- From the end of the conductor.

The energy reflected depends on the change in impedance. Oils have a low dielectric constant; water has a much higher dielectric constant. The reflection from the oil–water interface will be stronger than the reflection from the surface of the oil. However, a sophisticated microprocessor can detect both interfaces from a single probe.

There is a minimum distance required before the reflection can be dependably detected and accurately timed. Most guided wave radar probes have two dead zones, or blocking zones:

Top. This is analogous to the dead zone for noncontact radar.

Bottom. Because there is a reflection from the end of the probe, distinguishing this reflection from that of a level near the end of the probe is a problem, especially for materials with a low dielectric constant.

Buildups

Any change in the impedance causes some of the energy to be reflected. The following are capable of providing such an impedance change:

Thin coating. A thin coating on the probe will degrade the performance of the instrument but should not lead to a false level indication. The higher the dielectric constant of the coating, the more the degradation in performance.

Thick buildup or bridge. This gives a reflection much like the one from the surface of interest. Especially if the buildup or bridge has a high dielectric constant, the level measurement device might (probably?) base the level indication on the location of the buildup or bridge instead of the surface of interest.

Stratification. Stratification could provide a change in dielectric constant (and thus impedance) sufficient to reflect some of the energy. The effect would be most severe when the light material has a low dielectric constant and the heavy material has a high dielectric constant. In this context, stratification has the appearance of an interface.

It is very useful to have the capability to display the actual waveform sensed by the receiver when analyzing the source of such problems. This waveform can be plotted as the energy received as a function of the distance from the transmitter.

Probes

Three different types of probes are available; the choice is based primarily on the dielectric constant:

Single rod. The liquid or solid must have a dielectric constant of 10 or greater—that is, water or water-like materials. However, there are lots of sumps where such a probe can be used.

Dual rod. The liquid or solid must have a dielectric constant of 2 or greater. The length of this probe can be up to 50 ft (other types are limited to 20 ft or less).

Coax. These probes can detect interfaces from materials with a dielectric constant as low as 1.4. Such a probe could detect the level in a propane storage tank.

All types can be rigid or flexible (cables or ropes). Probes are available for high-pressure applications (5000 psig) and moderate temperatures (200 °C), although the pressure limit generally decreases with temperature. The single and dual rod probes can be coated for corrosion resistance.

Advantages and Disadvantages

The major advantages of guided wave radar for level measurement are as follows:

- No calibration required.
- Immune to vapor–space and material properties.
- No false reflections from tank internals.

The disadvantages are as follows:

- Intrusive.
- Less accurate than noncontact radar.
- Lower maximum range than noncontact radar.
- Buildups on probe, which can lead to problems.

A distinct advantage of guided wave radar is that no calibration is required. Time and distance are related by the speed of light, which is a universal constant.

Like RF capacitance devices, guided wave radar level measurement devices are intrusive. But unlike RF devices, guided wave radar is not affected by changes in the dielectric constant, as long as it remains above the minimum requirement for the selected probe design. RF capacitance is less affected by buildups, but otherwise guided wave radar offers distinct advantages.

As compared to noncontact radar, guided wave radar is less accurate, a typical statement being 0.1% of probe length but not less than 0.1 in (2.5 mm). However, this is adequate for many industrial applications.

Guided wave radar measurement devices are generally provided as two-wire, intrinsically safe versions. Four-wire versions are available, but do not provide enhanced performance.

4.8. NUCLEAR

Nuclear level measurement is based on the principle that a material absorbs gamma rays in proportion to its mass. As illustrated in Figure 4.25, a nuclear level gauge consists of two components mounted on opposite sides of the process vessel:

Source. This is the component of concern. The source emits gamma rays, which are very penetrating and are directed through the process in the direction of the detector. For level measurement, the source tends to be large.

Detector. The detector generates a signal that indicates the intensity of the incoming gamma rays. The most commonly known detector is the Geiger-Muller tube used in Geiger counters, but this type is not used in industrial nuclear level gauges.

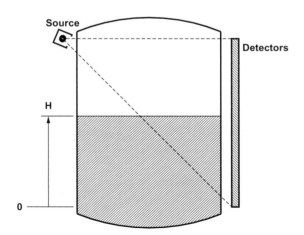

Figure 4.25. Level measurement by nuclear level gauge.

The least absorption occurs when the vessel is empty. As the vessel fills, more gamma rays are absorbed, so the intensity at the detector decreases. The most absorption occurs when the vessel is full.

Source

The size of a source is measured by the intensity of the radiation emitted. It is normally measured in millicuries; 1 g of radium emits 1 ci of radiation. The source in a nuclear level gauge may be as much as 20 ci, but most are far less. The materials used in the sources are generally one of the following, both of which are artificial isotopes:

Cesium 137. This is the most commonly used source. It has a half-life of 33 years. In most cases, a cesium 137 source will outlive the process.

Cobalt 60. This source outputs radiation with a higher energy level and thus is more penetrating. The larger the source, the more likely it will be cobalt 60. However, cobalt 60 has a half-life of 5.3 years. In early nuclear level gauges, this rapid decay rate necessitated more frequent calibrations. But with microprocessor-based technology, the decay of the source intensity can be taken into consideration by the software.

Sources are supplied in lead-encased steel containers with a shutter that can close off the source of gamma rays. Although strip sources were in use in the past, only point sources are used today. The radiation intensity at the detector is not a linear function of level, but microprocessors can take any geometry into account and output a signal that is linear with respect to the level.

Detectors

Two types of detectors are used in industrial nuclear level gauges:

Ion chamber. A long tube (up to 10 ft) with an electrode down its center is filled with an inert gas at high pressure. When a small voltage is applied to the electrode, the gamma rays ionize the gas and produce a very small current, which is proportional to the intensity of the gamma radiation striking the detector. Multiple ion chamber detectors and be installed to give a very wide span.

Scintillation counters. These are sometimes called *photon detectors.* When gamma rays strike the material within the scintillation counter, light particles known as photons are produced. A photo multiplier tube converts the photons into an electrical current, which is proportional to the intensity of the gamma radiation striking the detector. The usual length of a scintillation counter is 1 ft. They can be "stacked" to give measurement spans of 12 ft or more.

Scintillation counter detectors are more expensive, but they are about 10 times as sensitive as ion chamber detectors. This translates into a reduction in the source size by a factor of 10. In most level measurement applications, the design objective is to get the job done with the smallest possible source (least amount of radioactive material on-site).

Design

From vessel geometry, material properties, etc., the nuclear level gauge supplier does all design calculations. Most users need not be concerned with the details, but should appreciate issues such as the following:

- A major consideration is the radiation level to which people in the vicinity of the detector would be exposed. Obviously lower is better. The greater the sensitivity of the detector, the lower this radiation level will be.

- With the vessel empty, gamma rays are absorbed by the walls of the vessel, the vessel internals, and whatever else is between the source and the detector. If you know the vessel geometry, vessel material, etc., you can compute the fraction of the radiation that is absorbed by the vessel. As this fraction approaches 1.0, the size of the source increases rapidly. For example, if 90% is absorbed, the size of the source must be increased by a factor of 10.

- The radiation intensity decreases with the square of the distance from the source. Therefore, vessel size has a dramatic effect on the size of the source. One trick to decreasing this distance is to install the source in a well inside the vessel. But in the applications for which nuclear level gauges are considered, this is rarely feasible. Instead of traversing the main diameter of the vessel, the radiation beam can traverse a chord. However, this increases the percentage of the beam that is absorbed by the vessel walls.

- A change in level from 0% (empty) to 100% (full) must lead to a significant decrease in the intensity of the radiation at the detector. A decrease of at least 50% is desirable.

Safety

The first aspect of safety is that there can be no residual radioactivity in the materials flowing through process or within the process vessel itself. Gamma rays are penetrating and can be destructive, but they are permitted even when food-approval requirements are imposed. The risk of concern is personnel exposure. But let's make one thing clear: The proper procedures are well known. If these are followed, nuclear level gauges are completely safe. Most quickly put emphasis on the *if*. The fear is the consequence of a mistake. Nuclear level gauges are of particular concern because

- The radioactive source usually is a high-intensity source. The installation is designed so that the radiation levels at the detector are acceptably low so that even prolonged exposures are still within limits. But within an empty vessel, the radiation at locations closer to the source is far more intense.
- The process vessel is usually large enough for someone to physically get inside. The proper procedures call for the shutter on the radioactive source to be closed before entry (if you really want to be conservative, you could require that the source be physically removed before anyone enters the vessel).

There is no guarantee that the proper procedures will actually be followed in every instance. Unfortunately, the consequences of most concern do not surface immediately.

Regulations

Most nuclear level gauges require a Nuclear Regulatory Commission (NRC) fixed gauge license. More details are available in *Program-Specific Guidance about Fixed Gauge Licenses*,[2] which in turn references a variety of regulations. The regulations change from time to time, and some states get involved; today the issue of installations in foreign countries often arises. Fortunately, the suppliers of nuclear level gauges help their customers navigate this maze (or else they would have no customers).

Although the suppliers provide considerable assistance, the ultimate responsibility rests with the owner of the nuclear level gauge. Someone will have to become the resident *radiation expert*, formally known as the radiation safety officer (RSO). This is a formidable obstacle for the first nuclear gauge installation, especially in a small plant where no other installations are anticipated.

And then there is the paperwork. The nuclear device must be tracked from the instant it arrives to the instant it leaves. You pay to get these things; you pay to get rid of them. They have to be disposed of properly, usually by returning them to the supplier and paying a disposal fee.

Advantages and Disadvantages

The major advantages of nuclear level gauges are as follows:

- Completely external to the process.
- Unaffected by process temperature, pressure, etc.
- They work, even in very difficult applications.

The disadvantages are as follows:

- Radioactive materials on site.
- Regulatory issues.
- Consequences of mistakes can be serious.

When are nuclear level gauges installed? When nothing else will work, although occasionally when the process contents are so hazardous that the alternatives are even more risky.

And sometimes they are not installed even when there is no other alternative. Plant management takes a dim view of these things, observing that "We never pass up the opportunity to make a mistake." I have been told "No radioactive materials on site, end of discussion." At least the discussion was brief. This is an overreaction, mostly based on emotions, but it does happen. The reality is that the success rate with nuclear level gauges is very respectable, and mostly in very difficult applications. Nuclear level gauges deserve a better fate.

As the detector technology continues to advance, applications can be undertaken with ever-smaller sources. The smaller the source, the easier the licensing issues and the support issues. For example, leak tests are required only when the source exceeds a certain size. Will this ever be sufficient to moderate the concern? Probably not. Any personnel exposure, no matter how small, raises the potential for lengthy and costly adversarial proceedings.

4.9. A FEW OTHERS

Floats are used in a variety of configurations, two of which are presented here. In addition, *displacers* and *resistance tape* are also discussed.

Float and Tape

For large storage tanks, a float and tape system is one approach for measuring the tank level. As illustrated in Figure 4.26, the float rides on the liquid surface. Connected to this float is a measuring tape. A mechanism winds and unwinds the tape so as to maintain a constant tension on the tape.

In the past, the tape drove a visual indicator, usually mounted on the side of the tank so an individual could conveniently read it. Today, encoders are incorporated into the tape-winding mechanism so that its current position can be monitored remotely.

Such systems were commonly installed in early efforts to make a transition from manual tank level measurements to remote monitoring of tank levels. The objective was to replicate manual results, but the system is essentially an automated version of taking a manual measurement. Today, we are trying to eliminate measurement systems of a highly mechanical nature.

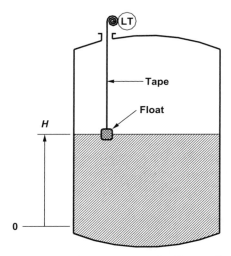

Figure 4.26. Level measurement by float and tape.

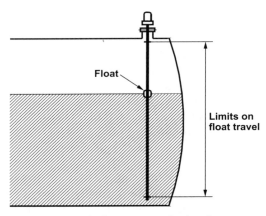

Figure 4.27. Magnetic float system for level measurement.

Magnetic Float Systems

In magnetic float systems, the float slides up or down a vertical rod, as illustrated in Figure 4.27. There are two technologies for detecting the position of the float.

With one approach, a chain of reed relay switches inside the rod are actuated by a magnet within the float. The switches provide a level resolution of about 10 mm. Tube lengths of up to 20 ft are practical. These units can tolerate process temperatures up to 350 °F and pressures up to 300 psig.

Another approach is to employ highly sensitive magnetostrictive position sensors. Applying a magnetic field to a magnetostrictive material causes stresses that change its physical dimensions, albeit by a very small amount. Transition elements such as iron, nickel, and cobalt are magnetostrictive. A wire made of magnetostrictive material exhibits a useful characteristic known as the Weidemann effect. When an axial magnetic field (such as from a permanent magnet) is applied at a point along the wire and current is passed through the wire, twisting occurs at the point at which the magnetic field is applied. If the current is applied as a short-duration pulse, the twisting action generates an ultrasonic pulse. The magnetostrictive wire serves as a waveguide, and the ultrasonic pulse travels back to the source of the current at the speed of sound in the waveguide material. This effect is the basis for a magnetostrictive position sensor.

In level measurement, the magnetostrictive wire is inside the rod that guides a float containing a permanent magnet. A magnetostrictive position sensor senses the position of the float. These units provide a level resolution of about 1 mm. The rod length is restricted to about 10 ft. Process temperature limit is about 170 °F; process pressure limit is about 150 psig.

Displacers

In the days of pneumatic instrumentation, displacers were very popular. Many are still installed in refineries. Although their popularity has declined, they continue to be available.

A displacer relies on the Archimedes buoyancy principle: The force exerted on the displacer is equal to the weight of liquid that is displaced. The displacer is maintained in a fixed position. As the liquid level rises, the buoyancy force on the displacer increases. To minimize motion, this force is applied to a torque arm. Unfortunately, a seal is required around the torque arm, which creates the potential for a leak.

Figure 4.28 shows a displacer level measurement product. Although not shown in the figure, isolation valves are normally inserted between the external chamber and the vessel. The practical measurement span for a displacer is limited to about 12 ft.

Resistance Tape

The basic construction of a resistance tape is a wire wound around a base conductor, thus forming a helix. As illustrated in Figure 4.29, this is then enclosed in an envelope of a flexible material. The construction is such that the wire does not normally make contact with the base conductor. But when submerged in liquid, the extra pressure forces the wire to make contact with the base conductor.

For level measurement, the resistance tape is lowered into the vessel from a top opening. The sensor then measures the resistance between terminals

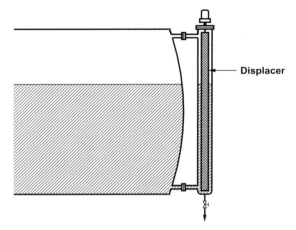

Figure 4.28. Displacer level measurement.

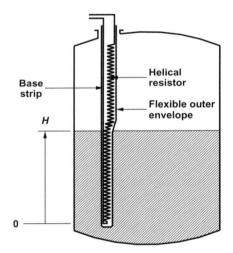

Figure 4.29. Level measurement by resistance tape.

attached to the wire and the base conductor. As the level in the vessel rises, the resistance decreases. The level can be sensed with a resolution of about ¼ in.

The resistance tape is flexible, so even long lengths can be installed from a top opening. If desired, a temperature sensor can be provided at the lower end of the tape. The resistance tape is intrusive, but there are no moving parts. The flexible envelope can be manufactured from corrosion-resistant materials. Coating and buildups do not affect performance, provided the buildup does not cause the wire to contact the base conductor.

4.10. LEVEL SWITCHES

The terms *active* and *passive* in the context of level switches are used as follows:

> *Passive.* The level switch does not require an external source of power. The following level switches are of this type:
> - Float.
> - Hydrostatic head (pressure).
>
> *Active.* An external source of power is required. The following level switches are of this type:
> - Capacitance.
> - Nuclear.
> - Ultrasonic.
> - Vibrating element.

Level switches are a common component of safety systems. To get high reliability, you keep it simple. This suggests that passive switches are preferable. But for level, the technology in the active switches offers advantages that frequently offset the disadvantages of requiring an external source of power. The most common level switches in recent plants are capacitance, ultrasonic, and vibrating elements.

Level Switch Installation

Although there are a few exceptions, the vessel level at which most level switches actuate is determined by their physical location. Many vessels must be equipped with more than one level switch. Some require a high-level switch and a low-level switch; some require a high-level switch and a high-high-level switch. Most plants will have at least one vessel with three (or more) level switches: low level, high level, and high-high level.

Figure 4.30 shows the two alternatives for installing level switches:

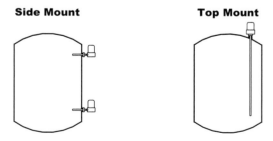

Figure 4.30. Level switch installations.

Side mount. Each level switch, located on the side of the vessel, requires a vessel connection, which is usually dedicated to the level switch. Of course, every connection provides a potential source of leaks. To gain access to the level switch, the vessel must be entirely drained (or at least to a level below that of the level switch).

Top mount. The level switches are located on a probe that is inserted from the top of the vessel. For buried tanks, this is the only viable approach. A single probe can provide multiple level switches. In many applications, the probe can be removed for servicing without draining the tank. However, structural issues arise for long probes, especially when inserted into agitated vessels.

Float

If you list all level switch applications, the float would surely be at the top of the list—some version of a float level switch is installed in most toilet bowls.

Industrial versions are available, such as shown in Figure 4.31. This one can be inserted horizontally through an opening for ½-in normal pipe threads. The following points apply:

- Their mechanical nature is a major disadvantage when used for many liquids encountered in process applications. Basically, they are limited to clean fluids.
- They are inexpensive, unless manufactured from special materials.
- Versions are available that can go up to about 200 °C and 300 psig.
- The one illustrated in Figure 4.31 must be installed horizontally. Versions for vertical installation are also available.

Pressure and Diaphragm

Any pressure switch is a potential candidate for a level switch. One approach is to locate the switch at or near the bottom of the vessel. Because the pressure at the switch location increases with vessel level, the switch can be adjusted

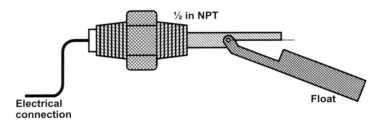

Figure 4.31. Horizontal float level switch.

to actuate at the appropriate level. However, the level at which the switch actuates is affected by liquid density, which in turn is affected by temperature and composition. Compensating for these increases the complexity and requires an external source of power. However, rarely is any ambiguity acceptable regarding the level at which the switch actuates.

In most requirements for a level switch, the physical location of the switch determines where the switch actuates. A diaphragm level switch is really a pressure switch that actuates on the increase in pressure when submerged in liquid. Most are mounted on a side connection to the vessel; however, versions that can be suspended in the vessel are available. They can be used up to about 100 °C and 75 psig, including slurry and liquid applications.

Capacitance

The capacitance level switch is an on–off version of the continuous capacitance probe. In fact, the probes for capacitance level switches are the same as the probes for continuous capacitance level measurements. The electronics module for the level switch can be coupled with a wide variety of probe designs (insulated or noninsulated, rigid or flexible, etc.) to meet almost any requirement. However, the probes for capacitance level switches are frequently short, rigid probes inserted from the side of the tank, such as the one in Figure 4.32.

The capacitance at which the switch actuates is adjustable. In some models, the deadband for reactuating is also adjustable. Adjustable time delays are also commonly provided. The electronics module is normally mounted directly

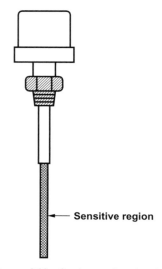

Sensitive region

Figure 4.32. Capitance level switch.

on the probe, as in the figure. The temperature limit for the electronics module is typically 80 °C, although the probe can withstand much higher temperatures (200 °C or more). Probes that can withstand pressures of 1500 psig are available.

Electronic modules that provide multiple level switches based on a single probe are also available. The probe configurations are very similar to those for continuous level measurements. But instead of providing a continuous output, the electronics module provides multiple contact outputs.

Ultrasonic

The probes for ultrasonic level switches always have a gap. On one side of the gap is the transmitter; on the other side of the gap is the receiver. Liquids are far better transmitters of sound than any gas. When the liquid level is below the gap, the receiver will not detect the sound. When liquid fills the gap, the receiver will detect the sound.

The model illustrated in Figure 4.33 has two gaps. Therefore, there are two level switches mounted on the same probe. The user can specify the locations of the gaps. This type of probe has to be inserted from the top of the vessel. However, probes with a single gap are normally inserted from the side of the vessel.

Temperature limits are typically −40° to 80 °C for the electronics. Somewhat higher temperatures are permitted on the probe, with typical limits being −40° to 120 °C. Probes with pressure limits as high as 1000 psig are available.

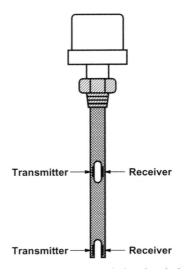

Figure 4.33. Ultrasonic level switch.

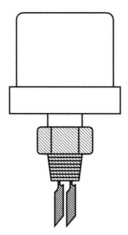

Figure 4.34. Vibrating element (tuning fork) level switch.

Vibrating Element

The vibrating element probes come in various shapes. The one shown in Figure 4.34 has two prongs. Some probes are cylindrical. The probe is vibrated at a frequency in the range of 100 to 500 Hz. When the probe is exposed to air, the amplitude of vibration is the greatest. When either liquids or solids surround the probe, the vibration is damped. This is sensed by the electronics and actuates the switch.

Vibrating element probes are most commonly applied to solids. Sometimes they are inserted horizontally, but they can be inserted vertically. In some applications involving solids, they work in one direction but not in the other. Working with solids is usually full of surprises.

Vibrating element probes are available for temperatures from −40° to 160°C. However, the electronics are limited to −40° to 60°C. Probes with pressure limits as high as 1500 psig are available.

Nuclear

Nuclear level switches have the same advantages and disadvantages as their continuous counterparts. They are totally external to the vessel and can even be outside jackets and insulation. But then there are all the problems associated with having radioactive materials on site.

Figure 4.35 illustrates using nuclear level switches for detecting high level and low level. The nuclear level switches can be installed with a single source and two detectors or can be installed with two sources and two detectors. Usually the decision depends on which configuration results in the least amount of radioactive material on the site.

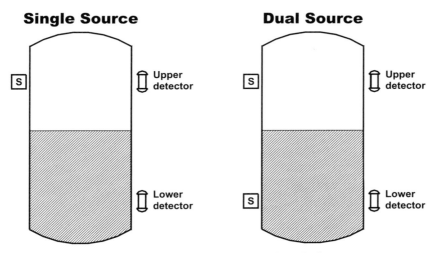

Figure 4.35. Nuclear level switch installation.

Nuclear level switches are extremely rare in industrial plants. Usually they are tolerated only when there is no other alternative.

4.11. INTERFACE

Water washing of an organic material is a common industrial operation. After the phases are thoroughly mixed, they are then allowed to separate in a decanter from which the phases can be separately removed. Some decanters are continuous; others are batch.

A decanter is only one example for which determining the location of an interface is important. Gravity is the driving force for separation, with the more dense (heavy) phase on the bottom and the less dense (light) phase on the top. Generally the aqueous phase will have a higher specific gravity than the organic phase, and consequently the aqueous phase is on the bottom.

Even in specialty chemical plants, the organic phase is commonly referred to as the *oil phase*, and the decanter as an *oil–water separator*. We shall use this terminology in the notation for the specific gravities (relative to water) of the two phases:

$G_{Aqueous}$. Specific gravity of the aqueous phase.

G_{Oil}. Specific gravity of the organic phase.

Differential Pressure—Dry Leg

The following approaches keep the piping to the upper connection free of liquids (a dry leg):

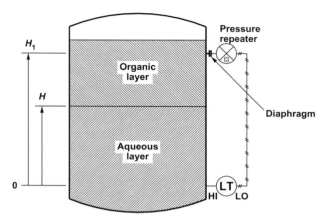

Figure 4.36. Interface level measurement using differential pressure and dry legs.

- Purging with a gas.
- Installing a pressure repeater (Fig. 4.36).
- Using two pressure transmitters and computing the differential pressure.

When the interface is at or below the lower connection, the hydrostatic head is H_1 times the gravity of the organic layer. When the interface is at or above the upper connection, the hydrostatic head is H_1 times the gravity of the aqueous layer. The configuration parameters for the differential pressure transmitter must be as follows:

Value	Indicated Value	Pressure at Transmitter, in H_2O
Lower-range value	0	$G_{Oil} H_1$
Upper-range value	H_1	$G_{Acqueous} H_1$
Span	H_1	$(G_{Acqueous} - G_{Oil}) H_1$

Differential Pressure—Wet Leg

The following approaches keep the piping to the upper connection full of a known liquid (a wet leg):

- Purging with a liquid.
- Installing capillary seal systems (Fig. 4.37).

On the process side, the hydrostatic head of the column of liquid between the connections is the same as for the dry leg configuration. On the instrument

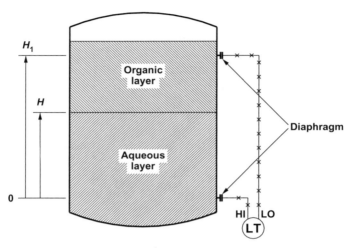

Figure 4.37. Interface level measurement using differential pressure and wet legs.

side, the hydrostatic head of the column of seal fluid is the height H_1 times the gravity G_{Seal} of the seal fluid. The configuration parameters for the differential pressure transmitter must be as follows:

Value	Indicated Value	Pressure at Transmitter, in H_2O
Lower-range value	0	$(G_{Oil} - G_{Seal})\, H_1$
Upper-range value	H_1	$(G_{Acqueous} - G_{Seal})\, H_1$
Span	H_1	$(G_{Acqueous} - G_{Oil})\, H_1$

Capacitance

In separating a heavy organic material from an aqueous phase, the difference in specific gravities can be quite small (for example, 0.2). This impairs the performance of the differential pressure approaches (and displacers and nuclear gauges). However, capacitance methods are not impaired.

The most common interface measurement application is when the upper phase is a nonconducting organic but the lower phase is aqueous and conducting. Because the aqueous phase is conducting, a coated probe (Fig. 4.38) is required. If the vessel is not full, the sensitive region of the probe should begin below the surface of the organic layer. If the surface of the organic layer is variable, the sensitive region should begin below the lowest possible value for the surface of the organic layer.

The aqueous phase will have a high dielectric constant but will be conductive. Therefore, the capacitance sensed by the probe within the aqueous layer

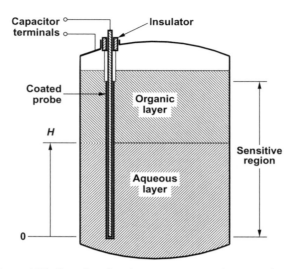

Figure 4.38. Interface level measurement using capacitance.

depends on the dielectric constant of the coating not on the dielectric constant of the aqueous phase. This turns out to be a major advantage, because the composition (and consequently the dielectric constant) of the aqueous phase is unlikely to be constant.

The dielectric constant of the organic phase contributes to the capacitance as sensed by the probe. As the interface rises, the contribution to the capacitance by the aqueous layer increases, but the contribution to the capacitance from the organic layer decreases. The sensitivity of the probe depends on the difference between these two contributions.

Ultrasonic

To sense the interface level, the ultrasonic transmitter–receiver must be wetted. There are two options, neither of which is very appealing:

- Locate the transmitter–receiver in the aqueous phase. In addition to the obvious issues with leaks, the aqueous phase is often corrosive.
- Install the transmitter–receiver in the organic phase, which usually means that either the vessel must be full or some support mechanism must be provided so that the it is immersed in the organic layer.

The ultrasonic approach also requires a well-defined interface. But when a rag layer is present (and it often is), the sound is largely dispersed instead of reflected.

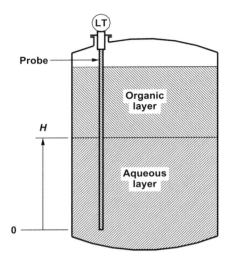

Figure 4.39. Interface level measurement using guided wave radar.

Guided Wave Radar

Guided wave radar obtains a reflection from any change in the impedance along the probe. In oil–water separators, the largest change in impedance is likely to be from the organic–water interface. For the application illustrated in Figure 4.39, there will be three reflections:

- From the surface of the organic layer.
- From the organic–water interface.
- From the end of the probe.

The reflection from the organic–water interface usually has a higher intensity than the remaining two. Some commercial systems can detect both the organic surface level and the interface level from a single probe.

The presence of a rag layer spreads the reflection from the interface. Detecting a broad reflection is more difficult than detecting a sharp reflection, but the intensity of the reflection from the organic–water interface is usually sufficient enough that even a broad reflection can be reliably detected. However, guided wave radar systems are affected by buildups on the probe in the vicinity of the interface.

Nuclear

The absorption of gamma rays by a material increases with the density of the material. Therefore, the higher the interface, the lower the intensity of the radiation at the detector. Nuclear interface measurement has one characteristic in common with differential pressure interface measurement. As the difference between the specific gravities of the two phases decreases, both the resolution and the accuracy suffer.

The advantages and disadvantages of the nuclear interface measurement are essentially the same as for the level measurement. Its major advantage is that all components of the measurement system are external to the process. Its major disadvantages are the presence of radioactive materials on site and all that this entails. If procedures are followed, they are safe. However, concern about the *if* is compounded by the fact that (*a*) the source tends to be large and (*b*) the process vessel is large enough for someone to physically get inside it. Like level measurement, nuclear systems are usually installed for interface measurement when there is no other alternative.

Float

Floats are available in a variety of different sizes, shapes, and weights. One can certainly obtain a float whose weight per unit volume (density) is lighter than the aqueous phase but heavier than the organic phase. Such a float would sink through the organic layer but not through the aqueous layer. It would remain at the interface, where it displaces a combination of organic and aqueous material whose weight is the same as the weight of the float.

The various float technologies that apply to liquid level measurement can then be applied to interface level measurement. In interface applications, the main concern is keeping the float clean. Buildups from either phase will interfere with the motion of the float, which soon sticks in a fixed position.

Displacer

The same displacer technology used in a liquid level measurement can be applied to interface measurement. The buoyant force on the displacer is the weight of liquid displaced by the displacer. The maximum buoyant force occurs when it is displacing the aqueous phase entirely. The minimum buoyant force occurs when it is displacing the organic phase entirely.

Displacers also have one characteristic in common with differential pressure (and nuclear) interface measurement. As the difference between the specific gravities of the two phases decreases, both the resolution and accuracy suffer.

Another serious concern in interface measurement applications is buildups on the displacer. Buildups from either phase basically change the effective weight-to-volume ratio for the displacer, leading to errors. For interface level measurement, the accuracy and resolution of a clean displacer is not very good, which makes even small amounts of buildups a problem.

4.12. DENSITY

Because the specific gravity is the density divided by the density of a reference material, *density measurement* and *specific gravity measurement* are really the same. For materials of known composition, the density or specific gravity can be inferred from temperature and pressure, both of which are much easier to measure than is density. But when composition is a variable, one is faced with either directly measuring the density or measuring the composition and inferring the density. Unless the composition is needed for other purposes, it is usually easier to directly measure the density.

Tank gauging is one application that often requires direct measurement of density. The liquid density is required to convert between the volume of material in a vessel and the mass of material in a vessel.

In many such applications, the density of a material flowing in a pipe must be measured. One example comes from the raw sugar industry. For solutions of sugar in water, the sugar concentration is inferred from a measurement of the density of the solution. Other similar applications in which concentration can be inferred from density are sodium hydroxide solutions and sulfuric acid solutions.

Differential Pressure

For measuring the density of a material in a tank, one approach is to use differential pressure. Figure 4.40 presents both a dry-leg arrangement and a wet-leg arrangement.

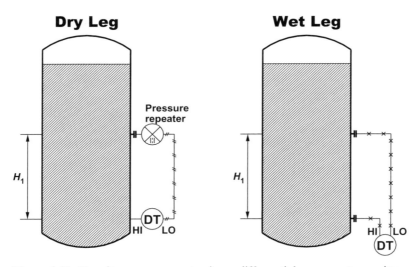

Figure 4.40. Density measurement using a differential pressure transmitter.

In a dry-leg arrangement, the differential pressure in inches of water (in H_2O) is the height H_1 of the column of liquid times the specific gravity G of the liquid. In the wet-leg arrangement, the liquid in the wet leg also contributes to the differential pressure. If the specific gravity of the seal fluid is G_{Seal}, this contribution is $H_1 G_{Seal}$. If the specific gravity is to be measured from G_{Min} to G_{Max}, the configuration parameters for the differential pressure transmitter with wet legs must be as follows:

Value	Indicated Value	Pressure at Transmitter, in H_2O
Lower-range value	G_{Min}	$(G_{Min} - G_{Seal}) H_1$
Upper-range value	G_{Max}	$(G_{Max} - G_{Seal}) H_1$
Span	$G_{Max} - G_{Min}$	$(G_{Max} - G_{Min}) H_1$

Displacer

The displacer is another level measurement technology that can be applied to measure density. Initially, one might be tempted to measure the density of a liquid in a tank using the same displacer configuration as used to measure liquid level. However, is the density of the liquid in the external chamber the same as the density of the liquid in the tank? To answer yes to this question, there needs to be some flow through the external chamber.

In specific gravity applications, displacers are normally used in a flow-through configuration. A compromise between sensitivity and response time is inevitable. To get maximum sensitivity, one needs as large a displacer as practical. But making the displacer larger increases the size of the chamber. At a given fluid flow, the larger chamber has a longer residence time.

Increasing the fluid flow raises questions about the effect of drag on the displacer. Figure 4.41 illustrates the use of a distribution ring to reduce the influence of drag. The liquid is admitted to the chamber containing the displacer through a distribution ring near the middle of the displacer. Fluid then flows in each direction, the objective being to minimize the net drag on the displacer due to fluid flow.

When measuring density in a flowing stream, the full flow usually does not pass through the density meter. Instead, a sample stream must be withdrawn and passed through the meter. This approach is also applied to measure density in tanks.

Vibrating Tubes

The natural frequency of vibration of a hollow metal tube depends of three factors:

- Length of the tube.
- Modulus of elasticity of the tube metal.
- Weight of the tube, including its contents.

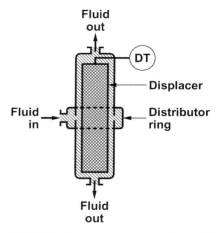

Figure 4.41. Density measurement using a displacer.

When a liquid or slurry is flowing through the tube, the weight of the tube depends on the density of the fluid. The manufacturing process includes custom calibration of the vibrational characteristics of each tube, including the influence of temperature. Dual tubes that are vibrated relative to each other provide immunity from vibrations within the process.

Commercial implementations that use this method use either a U-tube or a straight tube. Straight tubes offer lower pressure drop and are easier to clean, should this become necessary. Process temperatures may be in the range of −50° to 180°C. The meter can be installed horizontally, but a vertical configuration ensures that the tube is completely full of liquid. Accuracies of ±0.0001 g/cc are claimed. They are applicable to both liquids and gases.

Coriolis Mass Meter

For flow measurement, coriolis meters are routinely installed in specialty chemicals, pharmaceuticals, and similar industries. In addition to sensing the mass flow, these meters can also sense the temperature (required for compensating the effect of temperature on the vibrational characteristics of the tube). To measure the mass flow, the unit must sense certain vibrational characteristics of the tube. No additional measurements are required to determine the fluid density, so most units also provide a value for the fluid density. Viscosity requires additional information to be sensed, so this capability is provided as an option that must be purchased.

The coriolis principle is the basis for measuring the flow but not the density. The principle for measuring the density is that the frequency of vibration depends on the mass of the vibrating tube and thus the density of the material flowing through the tube. The principle is the same as for the vibrating tube density meters described previously. The term *coriolis density meter* is not really appropriate but is nevertheless used.

In plants that have coriolis mass meters installed elsewhere, they are a very attractive option even when the density is of interest but the mass flow is not.

Vibrating Element

The vibrating element may take the form of a tuning fork. The resonant frequency of the vibrating element depends on the following:

- Geometry (shape and dimensions) of the element.
- Modulus of elasticity of the metal.
- Density of the fluid surrounding the element.

As the density of the fluid surrounding the element increases, the resonant frequency of vibration decreases. The manufacturing process includes calibration of the vibrational characteristics of each element, including the dependence on temperature.

The tuning fork products can be applied to gases and liquids (viscosities up to 20,000 cP). The fluid temperature can be up to 200 °C, and the fluid pressure can be up to 3000 psig. Accuracies of ±0.001 g/cc are claimed. The vibrating element can be either inserted into a vessel or inserted directly into a pipeline.

For gas density, the vibrating element is sometimes a hollow cylinder. These products usually require a sample system to withdraw a sample stream from the process. The sample system usually provides for two standard gases that are used to establish the zero and span.

Nuclear

In nuclear measurement applications, the volume between the source and detector is known, and the nuclear device determines the amount of mass in this volume. This information is then used as follows:

Level measurement. The volume is partially filled, and the level within this volume is determined from the mass.

Density measurement. The volume is completely filled, and the density of this material is the mass divided by the volume.

The major issue in both cases is the presence of radioactive materials on site. However, the density measurement devices require much smaller sources, which greatly reduces the potential consequences of mistakes. The smaller source reduces the objections to nuclear density (and thickness) gauges but does not completely eliminate the issues associated with using radioactive materials.

The nuclear density measurement devices are excellent at determining the mean density of whatever material is within the meter. Unfortunately, this may not be the desired value. Consider a blender that mixes limestone and asphalt. The percent limestone in the mixture can be inferred from its density. Nuclear devices can easily tolerate the temperature of the mixture leaving the blender and can determine the density of this mixture. But there is a potential problem. Depending on the design of the blender, the potential exists for the blender to entrain some air into the mixture. The sensed density is the mean density of the limestone, asphalt, and air mixture. The percent limestone in the mixture cannot be inferred from this density.

LITERATURE CITED

1. Yaws, Carl L., *Chemical Properties Handbook*, McGraw-Hill, NewYork, 1999, table 8.
2. Nuclear Regulatory Commission, Program-Specific Guidance About Fixed Gauge Licenses, NUREG-1556, Vol. 4, Washington, 1998.

CHAPTER 5

Flow

In past years, the orifice meter was basically the default selection for a flow measurement device. However, this is no longer the case, especially in industries such as specialty chemicals and pharmaceuticals. A number of alternatives are now available that provide superior performance to that of the orifice meter.

This chapter covers the following technologies for industrial flow measurement:

Head-type flow meters. These include the orifice meter and venturi meter. New installations have declined, but lots of these meters are still in operation.

Coriolis meters. These measure mass flow, but they are expensive (especially in large line sizes). Their accuracy is superb, often surpassing that of load cells in industrial installations.

Magnetic flow meters. With no obstruction to flow, these work well on slurries and corrosive fluids.

Vortex shedding flow meters. Their cost is competitive with that of orifice meters, even in large line sizes.

Ultrasonic flow meters. These provide no obstruction to flow and are the very economical for large line sizes. The discussion includes two types: transit time and Doppler.

Thermal flow meters. These respond to changes in mass flow and are also very economical for large line sizes. The discussion includes two types: rate of heat loss and temperature rise.

Turbine meters. Bearing problems continue to be the major concern in liquid flow applications, but turbine meters are frequently encountered in gas flow measurements.

Others. Positive displacement meters; rotameters; target meters.

This chapter concludes by examining the options for flow switches.

Basic Process Measurements, by Cecil L. Smith
Copyright © 2009 by John Wiley & Sons, Inc.

5.1. MASS FLOW, VOLUMETRIC FLOW, AND VELOCITY

The flow through a pipe (or duct) can be expressed on any of the following bases:

Point velocity v *(m/min or ft/min).* Not all fluid particles in a flowing stream are moving at the same velocity. The point velocity is the velocity of the fluid at a specific point within the pipe. An *anemometer* is a device that can measure the velocity of the fluid at a point within a pipe or duct.

Average velocity V *(m/min or ft/min).* The average velocity is the average of the velocities of all particles passing a given point in a pipe or duct.

Volumetric flow Q *(m³/min or ft³/min).* Actual volume of fluid that passes a given point in a pipe per unit time. For gases, the common units are m³/min or ft³/min. For liquids, the common units are lpm or gal/min (gpm).

Mass flow W *(kg/min or lb$_m$/min).* Actual mass of fluid that passes a given point in a pipe per unit time.

The following relationships apply to a fluid flowing in a circular pipe:

$$Q = VA$$
$$W = \rho Q$$

where D = inside pipe diameter (m or ft); $A = \frac{1}{4}\pi D^2$ = inside cross-sectional area (m² or ft²); ρ = fluid density (kg/m³ or lb$_m$/ft³).

Volumetric Meters and Mass Meters

Industrial flow meters are categorized depending on what they directly sense:

Volumetric meters. Flow meters that sense the volumetric flow Q include the following:
- Head meters (orifice).
- Magnetic.
- Vortex shedding.
- Ultrasonic.
- Turbine.
- Rotameter.
- Positive displacement.
- Target.

Mass meters. Only two industrial flow meters directly sense the mass flow W:

- Coriolis (the term *mass meter* is often understood to be this type of flow meter).
- Thermal flow meter.

In most industrial applications, it is the mass flow that is of primary interest. The conversion of volumetric flow Q to the mass flow W is conceptually simple:

$$W = \rho Q$$

The problem is that the fluid density ρ depends on temperature, composition, and possibly pressure (usually insignificant for liquids, but always significant for gases). This introduces some uncertainty into the calculation.

Compressible and Incompressible

Compressible and *Incompressible* refer to the effect of pressure on the fluid density:

Compressible. The fluid density is affected by pressure. All gases are compressible. At very high pressures, liquids are also compressible, but the effect of pressure is much smaller than for gases.

Incompressible. The fluid density is not affected by pressure. In most industrial applications, liquids are considered to be incompressible. Gases are never treated as incompressible.

For gases, the effect of pressure is expressed by a relationship known as the equation of state. The simplest of these is the ideal gas law. The customary form relates the volume occupied by the gas to the moles of gas. But for flow applications, the form that relates the volumetric flow Q to the molar flow N is usually more convenient:

$$PQ = NRT$$

where P = absolute pressure (atm); R = gas law constant ($0.08205\,\text{atm-m}^3/$ kg-mol-K = $0.7302\,\text{atm-ft}^3/\text{lb-mol-}°\text{R}$); T = absolute temperature (K or °R); Q = volumetric flow rate (m^3/min or ft^3/min); N = molar flow rate (kg-mol/min or lb-mol/min).

Standard Conditions

When stating a volumetric flow, the temperature must always be stated. For gases, the pressure must also be stated. In many applications, the term *standard* conditions or *reference* conditions designates a specific temperature and pressure. This results in two bases for a volumetric flow:

Actual flow. This is the volumetric flow under flowing conditions.
Standard or reference flow. This is the volumetric flow at standard or reference conditions.

For gas flows, the fluid pressure and temperature are often sensed at or near the flow meter. Pressure and temperature compensation is the conversion of the actual volumetric flow sensed by the meter to a standard volumetric flow. The simplest equation to compensate for temperature and pressure is derived from the ideal gas law:

$$Q_{ref} = Q_{act} \frac{P_{act}}{P_{ref}} \frac{T_{ref}}{T_{act}}$$

where Q_{act} = actual gas flow; Q_{ref} = gas flow at reference conditions P_{act} = pressure (absolute) at flowing conditions; P_{ref} = pressure (absolute) at reference conditions; T_{act} = temperature (absolute) at flowing conditions; T_{ref} = temperature (absolute) at reference conditions.

When required, more accurate (and more complex) relationships are available.

Custody Transfer

Custody transfer encompasses flow measurement applications in which financial transactions are based on the measured values of the flow. The following issues arise in custody transfer applications:

- The accuracy of the flow measurement is of paramount importance.
- Provision must be made for passing a known flow through the meter to verify that its performance is within specifications.

In such applications, the general practices within the industry often dictate (or at least restrict) the choice of flow meter. For example, custody transfers of natural gas have traditionally been by head meters (orifice, venturi, etc.) that are individually calibrated. Various documents from the American Gas Association (AGA) specify in detail the design requirements (such as straight length of pipe upstream), how the differential pressure measurement is to be converted to flow, and so on.

Other meters have attributes that are desirable in such applications. Ultrasonic flow meters have a lower permanent pressure loss. Coriolis flow meters are more accurate. Through committees, organizations like the AGA consider the potential for applying such technology and issue documents that detail the procedures pertaining to the use of such meters for custody transfer.

Standards for Flow Meters

Standards for flow meters cover a variety of aspects, including the following:

Physical geometry. For example, the orifice for an orifice meter must be machined in a certain manner and to certain tolerances.

Design equations. Users rarely perform these calculations, but instead rely on computer programs provided by the manufacturers. The standards provide the basis for these programs.

Installation. For a given type of flow meter, the distance (if any) of straight pipe upstream required to obtain fully developed swirl-free flow depends on the source of the upstream disturbance and possibly some flow meter design parameters.

Standards for flow meters were first developed by the American Society of Mechanical Engineers (ASME), initially for head-type flow meters. Today, ASME standards are available for magnetic, vortex shedding, and many other flow meters. Flow meter standards from the ISO include procedures for calculating the uncertainty associated with a flow measurement. The API and AGA focus on topics of specific importance to their members, such as flow meter installations for custody transfer of natural gas.

5.2. STATIC PRESSURE AND FLUID VELOCITY

If you have a 3-in pipe, would you install a 3-in flow meter? Not necessarily. A common practice is to install a flow meter whose size is less than the size of the pipe. What are the incentives to install a 2-in flow meter in a 3-in line?

- For some types of flow meters, the cost of a 2-in meter is substantially less than the cost of a 3-in meter.
- For the 3-in flow meter, the manufacturer will state a maximum measurable flow and a turndown ratio. The minimum measurable flow is the maximum measurable flow divided by the turndown ratio. What can be done if flows less than the minimum must be measured? Consider installing a 2-in meter instead of a 3-in one.

For a given application, there is a maximum flow that the meter must be capable of measuring and a minimum flow that the meter must be capable of measuring. The flow meter size can be the same as the line size only when a meter of this size can measure both the maximum and the minimum flow required for the application. Problems are most likely to arise in regard to the minimum flow to be measured (undersize pipes are unusual). In such cases, a flow meter that is smaller than the line size is required. Especially for flow meters whose cost increases rapidly with line size, there is some incentive to install the smallest flow meter that can measure the maximum flow required for the application.

Area and Velocity

When the size of the flow meter is smaller than the line size, the following occurs:

- The fluid velocity within the flow meter is higher than the fluid velocity within the pipe.
- The static pressure within the flow meter is less than the static pressure within the pipe.

The lower pressure within the flow meter could lead to flashing within the flow meter. The net positive head within the flow meter must be positive under all conditions.

For the flow process shown in Figure 5.1, the volumetric flow Q is the same throughout. The velocities V_1 and V_2 are related to the areas A_1 and A_2, and to the pipe diameters D_1 and D_2, as follows:

$$\frac{V_2}{V_1} = \frac{A_1}{A_2} = \frac{D_1^2}{D_2^2}$$

Bernoulli Equation

Our concern is only for liquid flows, so we can use the following form of the Bernoulli equation, which applies to incompressible fluids:

$$\frac{P_1}{\rho} + \frac{V_1^2}{2} + gz_1 + W_s = \frac{P_2}{\rho} + \frac{V_2^2}{2} + gz_2$$

where P_i = static pressure at point i, (psia); V_i = fluid average velocity at point i (ft/sec); z_i = elevation at point i (ft); ρ = fluid density (lb$_m$/ft^3); W_s = shaft work (ft-lb$_f$/lb$_m$).

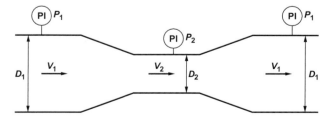

Figure 5.1. Effect of pipe diameter on fluid velocity.

Within a flow meter, there is no change in elevation ($z_1 = z_2$) and there is no shaft work ($W_s = 0$). The Bernouli equation can be expressed as follows:

$$P_1 - P_2 = \frac{\rho(V_2^2 - V_1^2)}{2}$$

The volumetric flow Q through the flow meter must be the same as the volumetric flow through the pipe. The flow velocities are related to the diameters as follows:

$$V_i = \frac{Q}{A_i} = \frac{4Q}{\pi D_i^2}$$

Substituting into the Bernoulli equation gives the following expression for the pressure within the flow meter:

$$P_2 = P_1 - \frac{8\rho Q^2}{\pi^2} \left[\frac{1}{D_2^4} - \frac{1}{D_1^4} \right]$$

Velocity Head

For flow through horizontal pipes, the Bernoulli equation states that

$$\frac{P}{\rho} + \frac{V^2}{2} = \text{constant}$$

Any change in the velocity V is accompanied by a change in the static pressure P. Hence the term *velocity head*, which suggests that the kinetic energy associated with the motion of the fluid is equivalent to a certain static pressure head.

The term *head meter* applies to any meter that senses the change in pressure that results from some change in the fluid velocity. The term *velocity change flow meter* is occasionally used and is quite appropriate. The following list contains the head meters that are considered for industrial flow measurement:

- Orifice meter.
- Venturi meter.
- Flow tube.
- Flow nozzle.
- Elbow taps.
- Pitot tube.
- Averaging Pitot tube.

Of all such meters, the orifice meter is by far the most common.

5.3. FLASHING AND CAVITATION

For a given substance at a given temperature, there is a unique pressure at which the liquid phase and vapor phase would coexist. This is the basis for the following:

- The pressure is the vapor pressure of the substance at that temperature.
- The temperature is the boiling point of the substance at that pressure.

Figure 5.2 shows a plot of the vapor pressure of water as a function of temperature. Note that at 100 °C, the vapor pressure is 1 atm. When pressure is plotted as a function of temperature (as in the figure), the relationship is known as the *vapor pressure curve*. When the temperature is plotted as a function of pressure, the relationship is known as the *boiling point curve*. The data points on the vapor pressure curve and the boiling point curve are exactly the same.

Vapor Pressure Correlations

The vapor pressure at a given temperature can be computed via the Antoine equation or extensions thereof. The Antoine equation contains three coefficients:

$$\log_{10}(P^\circ) = A + \frac{B}{T+C}$$

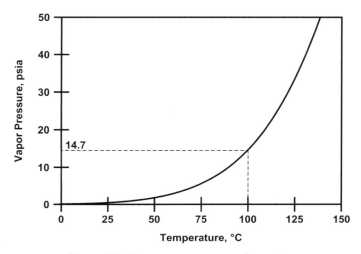

Figure 5.2. Vapor pressure curve for water.

where P^o = vapor pressure (mm Hg); T = absolute temperature (K); A, B, and C = coefficients.

Sometimes the temperature is in °C, the only consequence being that the coefficient C is larger by 273.15.

Various extensions and/or modifications to the Antoine equation have been formulated, with the objectives of providing more accurate values for vapor pressure and for covering a wider range of temperatures. One alternate equation contains five coefficients:

$$\log_{10}(P^o) = A + \frac{B}{T} + C\log_{10}(T) + DT + ET^2$$

where P^o = vapor pressure (mm Hg); T = absolute temperature (K); A, B, C, D, and E = coefficients.

Another important issue is obtaining values for the coefficients in the equation for the vapor pressure. Yaws[1] provides values for these coefficients for many organic and inorganic compounds.

Mixtures

To calculate the vapor pressure of the mixture, the following must be known:

Composition of the liquid. Usually this is a mole fraction.

Liquid temperature. This permits the vapor pressure of each component to be computed.

Activity coefficients. These depend on the molecular behavior of the components, specifically, on the tendency of molecules of one component to attract (or repel) the molecules of other components. When no such effects exist, the liquid phase behavior is said to be ideal, and the activity coefficient for every component is 1.0.

The vapor pressure of a mixture is calculated using the following equation:

$$P^o = \sum_{i=1}^{N} \gamma_i P_i^o x_i$$

where P^o = vapor pressure of mixture (mm Hg); P_i^o = vapor pressure of component i (mm Hg); x_i = mole fraction of component i in the mixture; γ_i = activity coefficient for component i; N = number of components in the mixture.

When ideal liquid phase behavior can be assumed, the vapor pressure of the mixture is relatively easy to calculate. But is this assumption valid? Only those familiar with the process technology can answer this question. If the liquid phase behavior is not ideal, these same people must either provide values for the activity coefficients or provide a value for the vapor pressure of the mixture at the temperature of the fluid within the meter.

Flashing

Suppose you have a liquid at a certain temperature and pressure. What would cause the liquid to vaporize?

> *Increase the temperature (add heat).* When the temperature attains the boiling point, the liquid will vaporize, or boil.
>
> *Decrease the pressure.* When the pressure attains the vapor pressure of the mixture, the liquid will vaporize, or flash.

Figure 5.3 illustrates both of these on the boiling point curve for water.

No flow meter adds significant heat to the fluid, so increasing the temperature does not happen. But low pressure can occur in many flow meters. In such meters, flashing is a potential problem. The result is

- Excessive measurement noise.
- Inaccurate measured value for the flow.

When the drop in pressure is immediately followed by an increase in pressure, some liquid vaporizes (flashes) and then quickly condenses. This is referred to as *cavitation*. Cavitation can damage the internals of a flow meter much like it damages the internals of control valves.

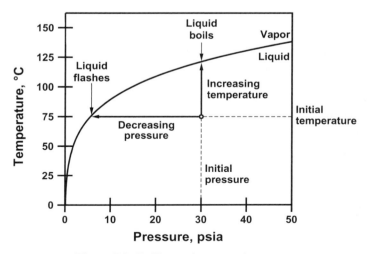

Figure 5.3. Boiling point curve for water.

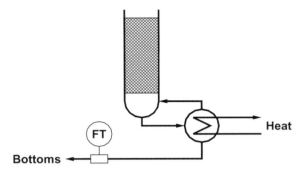

Figure 5.4. Measuring bottoms flow from a tower.

Net Positive Head

The net positive head is the static pressure less the vapor pressure of the liquid. Basically, this is the maximum decrease in pressure than can be permitted and still avoid flashing.

One is occasionally faced with measuring the flow of a stream whose net positive head is essentially zero. For example, the bottoms stream from the distillation tower shown in Figure 5.4 will have a net positive head of essentially zero. The liquid (reflux) from the lower packed section is partially vaporized by adding heat via an exchanger usually referred to as a *reboiler*. The remaining liquid exits as the bottoms stream. Because the liquid at the bottom of the tower is boiling, the only contribution to the net positive head is whatever hydrostatic head is provided by that liquid. In such applications, one has three options:

- Install a flow meter with no pressure drop, such as a magnetic flow meter or an ultrasonic flow meter.
- In some applications, the bottoms stream is immediately cooled in another heat exchanger. Install the flow meter downstream of this exchanger.
- In some applications, the bottoms stream is pumped to the next unit operation (the net positive head is usually an issue for this pump). Install the flow meter downstream of this pump.

5.4. FLUID DYNAMICS

The viscosity of a fluid characterizes its resistance to flow. "Thin" materials flow freely, and have a low viscosity. "Thick" materials flow slowly, and have a high viscosity. There are two measures of viscosity:

Dynamic viscosity μ. The common units are centipoise (cP):

   1 Poise = 1 Newton·sec/m² = 1 Pascal·sec

   1 cP = 0.01 Newton·sec/m² = 0.01 Pascal·sec = 2.42 lb$_m$/ft·hr

   The viscosity of water at 68 °F is 1.0 cP. The viscosity of air at 68 °F and atmospheric pressure is 0.018 cP.

Kinematic viscosity ν. The common units are centistokes (cSt):

   1 stoke = 1 m²/sec

   1 cSt = 0.01 m²/sec = 387.5 ft²/hr

The dynamic viscosity and the kinematic viscosity are related as follows:

$$\nu = \frac{\mu}{\rho}$$

where ρ = fluid density (kg/m³ or lb$_m$/ft³).

   The effect of temperature on viscosity is different for liquids and gases:

Liquids. Viscosity decreases with temperature.

Gases. Viscosity increases with temperature.

Velocity Profile

The velocity of the fluid at a specific location in the pipe is the point velocity. The velocity profile is a plot of the point velocity as a function of location within the pipe. The nature of the velocity profile depends on the flow region:

Laminar flow. Also known as viscous or streamline flow, the layers of fluid slide in the direction of flow with no mixing in the radial direction. As illustrated in Figure 5.5, the velocity profile is parabolic.

Turbulent flow. Mixing by eddy motion occurs between the layers of fluid. As shown in the figure, the velocity profile is essentially flat, dropping off only in close proximity to the pipe wall.

Transition. Laminar flow, turbulent flow, or some combination thereof may exist.

Reynolds Number

To determine the flow region, a dimensionless number known as the Reynolds number must be computed. For circular pipes the Reynolds number is defined as follows (expressions are available for other pipe geometries, but will not be presented herein):

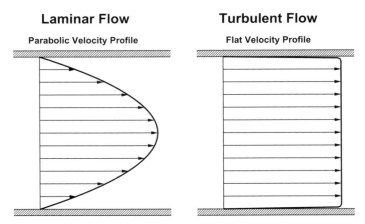

Figure 5.5. Velocity profiles.

$$N_{Re} = \frac{DV\rho}{\mu} = \frac{DQ\rho}{A\mu} = \frac{4Q\rho}{\pi D\mu} = \frac{4W}{\pi D\mu}$$

where V = average velocity (m/sec or ft/sec); Q = volumetric flow (m³/sec or ft³/sec); W = mass flow (kg/sec or lb$_m$/sec); D = pipe inside diameter (m or ft); $A = \frac{1}{4}\pi D^2$ = pipe inside area (m² or ft²); ρ = fluid density (kg/m³ or lb$_m$/ft³); μ = dynamic viscosity (kg/m·sec or lb$_m$/ft·sec).
The Reynolds numbers for the flow regions are as follows:

Laminar flow: $N_{Re} \leq 2100$.
Transition: $2100 < N_{Re} < 3500$.
Turbulent flow: $N_{Re} \geq 3500$.

Factors such as pipe roughness affect the transition to turbulent flow. When turbulent flow is required, some use a Reynolds number of 10,000 for the onset of well-established turbulent flow.
Some flow meters are affected by the Reynolds number; others are not. For example, the Reynolds number is crucial for the vortex-shedding meter. This meter relies on the von Karman effect, which occurs only in turbulent flow. In laminar flow, the output of this meter is zero.

Flow Disturbances

Fittings (elbows, tees, reducers, expanders, etc.) and valves introduce disturbances in the form of turbulence or swirls into a flow stream. The effect of these on the various types of flow meters is as follows:

Type of Meter	Effect
Head meters (orifice)	Significant
Vortex shedding	Significant
Ultrasonic	Significant
Turbine	Significant
Thermal	Significant
Target	Significant
Coriolis	Little or none
Magnetic	Little or none
Rotameter	Little or none
Positive displacement	Little or none

To deliver their stated performance, most types of flow meters require a well-established velocity profile that is free of swirls. Both aspects have been quantified in the standard ISO 5167-1:[2]

Swirl free. At every point over a pipe cross-section, the point direction of fluid flow is at an angle that does not exceed $2°$ from the axial direction.

Fully developed flow. The velocity profile is well-established if, at every point over a pipe cross-section, the point velocity is within 5% of what would exist at the end of a very long straight pipe (pipe length L exceeds $100D$, where D is the inside pipe diameter).

To achieve swirl-free, fully developed flow requires a minimum length of straight pipe upstream and downstream of the flow measurement device.

Straight Length of Pipe

The following data are from ISO 5167-2 (orifice)[3] and ISO-5167-4 (venturi)[4] and are for $\beta = 0.5$ (β is ratio of orifice or throat diameter to pipe inside diameter) and a fully open gate valve:

Type of Meter	Source of Disturbance	Additional Uncertainty	Length Upstream	Length Downstream
Orifice	90° Elbow	Zero	$22D$	$6D$
		0.5%	$9D$	$3D$
	Gate valve	Zero	$12D$	$6D$
		0.5%	$6D$	$3D$
Venturi	90° Elbow	Zero	$9D$	$4D$
		0.5%	$3D$	$4D$
	Gate valve	Zero	$3.5D$	$4D$
		0.5%	$2.5D$	$4D$

The primary source for recommendations is the standard applicable to the specific type of meter. Recommendations from the manufacturers generally adhere to the standards, except when proprietary technology is involved. The recommended lengths depend on the following:

Meter type. The previous values are for orifice and venturi meters only.

Meter design parameters. The values presented previously are for $\beta = 0.5$. For an orifice meter with $\beta = 0.6$, an upstream length of $42\,D$ is required for a $90°$ elbow and zero additional uncertainty. However, the length does not increase dramatically for even higher values of β.

Source of flow disturbance. It is not just the flow disturbance immediately upstream of the meter that is important. For example, an orifice meter $(\beta = 0.5)$ with a gate valve at $15\,D$ upstream and a $90°$ elbow at an additional $5\,D$ upstream (a total of $20\,D$) does not meet the above recommendations for zero additional uncertainty. The gate valve is Okay, but not the elbow.

Additional uncertainty. The current ISO standards pay great attention to uncertainty. When larger uncertainties are acceptable, shorter lengths can be used.

Flow Straighteners

Also called *vanes*, flow straighteners eliminate or significantly reduce the swirl. However, they do not necessarily produce the velocity profile for fully developed flow. Figure 5.6 illustrates three common types.

The design parameters for flow straighteners are also specified in the standards. The following are summarized from ISO 5167-1:[2]

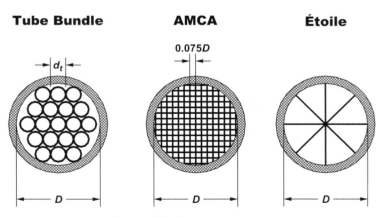

Figure 5.6. Flow straighteners.

Tube bundle. The bundle must consist of at least 19 tubes installed parallel to each other and with the pipe axis. The tube diameter d_t must not exceed $0.2\,D$. The tube length L must be at least 10 times the tube diameter, but in the range $2\,D \le L \le 3\,D$ (preferably close to $2\,D$). The pressure drop is approximately 0.75 velocity heads.

AMCA straightener. The vanes constitute a square mesh. The width of the squares is $0.075\,D$. The length of the vanes is $0.45\,D$. The pressure drop is approximately 0.25 velocity heads.

Étoile straightener. There are eight vanes at equal angular spacing. The length of the vanes is $2\,D$. The pressure drop is approximately 0.25 velocity heads.

For each type, the vanes should be constructed with the thinnest material that provides sufficient structural rigidity.

The installation of a flow straightener between a $90°$ elbow and an orifice meter is illustrated in Figure 5.7. For orifice meters, ISO 5167-2[3] provides the following statements for using a 19-tube bundle flow straightener:

- Use only for orifice meters with $\beta \le 0.67$.
- The distance L_s between the downstream face of the fitting and the orifice must be at least $30\,D$ $(L_s \ge 30\,D)$.
- The distance L_f between the downstream end of the tube bundle and the orifice must be $13\,D \pm 0.25\,D$.

If these conditions are met, the 19-tube bundle flow straightener can be used downstream of any fitting. However, these are generally conservative and are for zero additional uncertainty. Under certain situations, the distance L_s can be as short as $18\,D$. The applicable distances are provided by a table in ISO 5167-2[3].

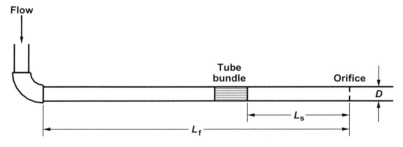

Figure 5.7. Flow straightener downstream of a $90°$ elbow.

Flow Conditioners

In addition to reducing the swirl, flow conditioners also redistribute the velocity profile, thus producing the conditions for swirl-free fully developed flow. As compared to vanes, their length is shorter, they are easier to fabricate, and they are more robust. On the negative side, the pressure drop is larger.

The designs of flow conditioners are generally proprietary. The following list is extracted from ISO 5167-1:[2]

- Gallagher.
- K-Lab NOVA.
- NEL (Spearman).
- Sprenkle.
- Zanker.

ISO 5167-2[3] includes the following statements for using a Zanker flow conditioner with an orifice meter:

- Use only if $\beta \leq 0.67$.
- Distance L_f to nearest upstream fitting must be $17D$ or greater.
- The distance L_s from the downstream face of the flow conditioner to the orifice plate must be $7.5D \leq L_s \leq L_f - 8.5D$.

Similar, but not identical, statements are made for the Gallagher and K-Lab NOVA flow conditioners. ISO 5167-1 provides a compliance test for a flow conditioner; any flow conditioner that has passed the test can be installed.

5.5. FLOW METER APPLICATION DATA

Some operating and engineering companies have prepared a form for specifying the application data for a flow meter. If you work for one of these, then certainly use its form. Some meter manufacturers have also prepared a form, usually tailored to a specific type of meter (magnetic, thermal, Doppler, etc.). The information required by the form is sufficient for most applications, but occasionally one must supplement the required data with other information that is relevant to an application.

For measuring the flow of gases or liquids, the application data falls into four categories:

Measurement parameters. This designates the required performance, such as measurement range (and units) and accuracy.

Fluid properties. This describes the nature of the fluid (liquid, gas, or two-phase) whose flow is to be measured.

Process piping. This describes the flow system into which the flow meter is to be installed.

Electronics. This includes issues such as the hazardous area classification that pertain to the electronics.

Measurement Parameters

Measurement parameters specify the capabilities required of the flow meter, including the following data:

Flow rate. In addition to the nominal value, the maximum and the minimum flows to be measured are required. Alternatively, the maximum flow and a turndown ratio could be specified.

Mass vs. volumetric flow. For liquids, is mass flow to be directly measured or is a volumetric flow measurement acceptable? For gases, is pressure and temperature compensation required?

Required performance. Accuracy is the most common specified performance parameter, but measurement uncertainty is an alternative specification. In some applications, however, repeatability is more appropriate.

Measurement range and units. For most flow meters, the lower-range value is zero and the upper-range value is the maximum measurable flow. The units do not necessarily suggest type of meter (mass flow units may be specified for volumetric meters, implying that constant density can be assumed).

Process conditions. These must convey the range of process conditions to which the flow meter will be exposed. Usually this takes the form of nominal, maximum, and minimum values for process temperature and pressure. But in practice, would the maximum process temperature occur at minimum process pressure?

Fluid Properties

Fluid properties specify the nature of the fluid flowing through the flow meter, including the following data:

Name of fluid. Manufacturers of flow meters understand terms such as *98% nitric acid*. But when uncommon chemicals are involved, the operating company probably knows more about the fluid than the manufacturer.

Fluid composition. For mixtures, a typical composition analysis should be stated. For those anticipated to lead to problems, materials present in trace amounts should be noted.

Suspended and entrained materials. This basically specifies the nature of any two-phase flow. This includes solids in either a gas or liquid, gas entrained in a liquid, or liquid droplets in a gas.

Fluid properties. Density and viscosity are generally required at nominal operating conditions. When the meter must operate over a range, data should be given at the operating limits. Where flashing is a possibility, vapor pressures must also be stated.

Materials of construction. For well-known materials, such as 98% nitric acid, the manufacturer can be expected to be knowledgeable. But for uncommon chemicals, the desired or acceptable materials of construction should be stated.

Process Piping

Process piping parameters specify the physical environment in which the flow meter will be installed, including the following data:

Pipe size and schedule. For standard circular pipes, the nominal pipe size and schedule is sufficient. Otherwise, the physical dimensions of the pipe or duct must be described in detail.

Material of construction. The material from which the pipe is constructed must be described, including any lining.

Preferred pipe connection. This is either screwed or flanged.

Sketch of installation. The anticipated piping upstream and downstream of the flow meter should be sketched, including lengths and fittings. This is especially important for flow meters that require straight lengths of pipe upstream and downstream.

Electronics

There are a number of issues that pertain to the electronics, including its required capabilities and the conditions to which it will be exposed. The specifications listed here pertain to the process conditions. If the electronics must be mounted remotely, the manufacturer can so state and specify the conditions required at that location.

Output signal. This is usually either current loop or fieldbus. If fieldbus, the specific type of fieldbus interface must be stated. But occasionally, a serial communications interface is required.

Area classification. This designation is crucial. If the location where the flow meter is to be installed is not classified, this must be specifically stated.

Intrinsically safe. This may be mandatory, may be preferred, may be optional, or may not be required (which means explosion-proof or purged enclosures will be stated, if appropriate).

Enclosure. Normally this is Nema 4 or 4X, unless an explosion-proof or purged enclosure is required.

Source of power. The options are usually 24 V DC, 110 V AC, or 220 V AC.

Local indication. Is the measured value to be indicated at or near the location of the flow meter?

5.6. ORIFICE METER

Most orifice meters sense the pressure drop resulting from a concentric orifice that is inserted into the pipe. As illustrated in Figure 5.8, the smaller diameter at the orifice plate causes the fluid velocity to increase, which causes the pressure to decrease. The lowest pressure (maximum pressure drop) occurs downstream of the orifice plate at a point known as the vena contracta. Although most of the pressure drop is recovered, there is a permanent pressure loss.

At one time, the approach to selecting a flow meter was to install an orifice meter unless there was a specific reason not to do so. This is no longer the case, and the share of such new installations is steadily declining. There is still a large installed base of orifice meters, but in industries such as specialty chemicals, very few orifice meters are being installed in new plants.

Extensive data are available on the performance of an orifice meter, much of it dating back to the 1960s or even earlier. These relationships are now incorporated into PC-based programs that perform the orifice meter design calculations. The discussion herein assumes that such a program will be used, so a detailed presentation is not warranted.

Orifice Plate Types

Figure 5.9 illustrates three types of concentric orifice plates. Detailed specifications are available as to the physical dimensions (thicknesses, radii, etc.) of each type. Their major features are summarized as follows:

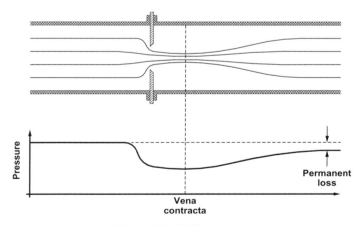

Figure 5.8. Orifice meter.

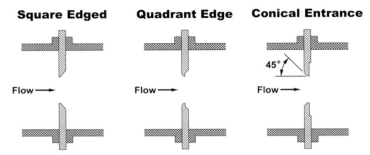

Figure 5.9. Types of concentric orifice plates.

Square edged or sharp edged. The square-edged concentric orifice gives the best accuracy at high Reynolds numbers. Consequently, this is the most common type.

Quadrant edge or quarter circle. This type is normally used for fluids with a high viscosity, which means that the Reynolds number will be lower.

Conical entrance. This type can be used at even lower Reynolds numbers than the quadrant edge.

Concentric orifices have a circular bore at the center of the pipe and provide the best accuracy. Eccentric orifices have a circular bore tangent to an edge of the pipe, normally at the bottom of the pipe to accommodate liquids with small amounts of solids or gases with small amounts of liquids. Segmental orifices have a noncircular bore and can accommodate higher amounts of entrained materials.

Pressure Tap Locations

There are five possibilities for the pressure tap locations:

Type	Location of Upstream Tap	Location of Downstream Tap
Vena-contracta	$0.5–2D$	At vena-contracta
Radius	D	$0.5D$
Flange	1 in	1 in
Corner	At orifice plate	At orifice plate
Pipe	$2.5D$	$8D$

The considerations include the following:

Vena-contracta taps. These sense the maximum differential pressure. But if the size of the orifice plate is changed, the location of the vena-contracta taps must also change.

Radius taps. These come close to the location of the vena-contracta, but have the advantage of being at a fixed location.

Flange taps. These can be incorporated into a special flange known as an orifice flange, thus avoiding taps into the pipe.

Corner taps. These are usually incorporated directly into the orifice plate itself.

Pipe taps. These basically sense the permanent pressure loss.

The usual practice is to use flange taps for small pipes and radius taps for large pipes.

Relationship for Orifice Meter

The relationships for the orifice meter are based on the Bernoulli equation. However, a number of correction factors, based largely on empirical data, are also required.

For a given orifice plate in a specific application, the relationships can usually be simplified to the following expression:

$$Q = C_o \left[\frac{\Delta P}{G} \right]^{1/2}$$

where Q = volumetric flow; ΔP = pressure drop at pressure tap locations; G = fluid specific gravity; C_o = orifice coefficient.

Most industrial applications are based on this equation, the main exception being custody transfer applications for which additional factors are incorporated to improve the accuracy. The value of the orifice coefficient C_o is determined as part of the design calculations. The units on C_o must be consistent with the units for the volumetric flow Q and the differential pressure ΔP.

The accuracy of orifice meter flow measurements is in the range of 1 to 2% for a clean orifice plate with no wear or corrosion. To obtain better accuracies (such as in custody transfer applications), orifice meters must be individually calibrated by passing a known flow through the meter.

Design Calculations

PC-based programs for performing the orifice plate design calculations are available from a variety of sources. Before PCs, designing an orifice plate required detailed familiarity with various relationships and correction factors. However, the PC-based orifice meter design programs are now universally used for such calculations. The input data to these programs include the following:

- Pipe size.
- Volumetric flow.

- Fluid properties.
- Type of orifice plate.
- Pressure tap locations.

The results of the design calculations encompass the following:

Orifice plate size. This is expressed as the β ratio, which is the ratio of the diameter of the orifice plate opening to the inside diameter of the pipe.

Orifice coefficient C_o. This relates the volumetric flow to the pressure drop at the pressure tap locations.

Permanent pressure loss. The available pressure drop for the flow system containing the orifice meter must provide for this.

Maximum pressure drop. The net positive head must be such that no flashing occurs within the meter.

Sonic velocity. For gas applications, the maximum velocity within the orifice meter cannot exceed the sonic velocity.

Flow vs. Differential Pressure

Figure 5.10 presents the generic relationship for flow as a function of differential pressure for an orifice meter. The maximum differential pressure ΔP_{max} is the upper-range value for the differential pressure transmitter. The maximum flow F_{max} is computed from ΔP_{max} and the orifice coefficient C_o. In terms of ΔP_{max} and F_{max}, the relationship for the orifice meter is expressed as follows:

$$\frac{F}{F_{max}} = \left[\frac{\Delta P}{\Delta P_{max}} \right]^{1/2}$$

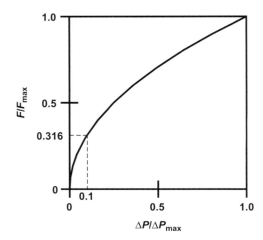

Figure 5.10. Flow vs. differential pressure for an orifice meter.

This is a nonlinear relationship. Linearization requires taking the square root of the differential pressure measurement. Most digital control systems provide this option as part of the input processing computations. However, the trend is to perform the square root extraction within the differential pressure measurement device. When this approach is taken, the measurement device is often said to "transmit in flow units"—that is, its output is linear in regard to flow. Without the square root extraction, the output is linear in regard to differential pressure.

Turndown Ratio

Square root extraction provides a linear output with respect to flow. However, other consequences of the nonlinear relationship remain. One of these is its effect on the turndown ratio.

To understand the problem with the turndown ratio, suppose a pressure transmitter has a turndown ration of 10:1. That is, it can accurately sense pressures down to $0.1\,\Delta P_{max}$. What is the corresponding flow? The flow is

$$\sqrt{(0.1)}\,F_{max} = 0.316 F_{max}$$

A turndown ratio of 10:1 on differential pressure translates into a turndown ratio of approximately 3:1 on flow. Differential pressure transmitters are usually capable of better than a 10:1 turndown ratio, so some will argue that the turndown ratio for an orifice meter is as much as 5:1. To obtain a 10:1 turndown ratio for the flow would require a 100:1 turndown ratio on the differential pressure measurement. Smart differential pressure transmitters provide such a turndown ratio, but there is another problem.

The other problem pertains to noise. At low flows, the square root extraction is amplifying any noise in the differential pressure measurement. For $\Delta P < \frac{1}{4}\Delta P_{max}$ or $F < \frac{1}{2}F_{max}$, the noise is being amplified. This is another obstacle to attaining a 10:1 turndown ratio. For $\Delta P = 0.01\,\Delta P_{max}$ or $F = 0.1\,F_{max}$, the noise is being amplified by a factor of 5.

Installation

The issues pertaining to the installation of the differential pressure transmitter are summarized as follows:

Service	Location of Pressure Taps	Location of ΔP Transmitter	Connecting Lines
Gas	Top of pipe	Above pipe	Sloped so liquid drains down
Liquid	Side of pipe	Below pipe	Sloped so gas escapes up
Condensing vapor	Side of pipe	Below pipe	Filled with condensate

These are essentially the same as for pressure transmitters in these services.

The various orifice meter relationships assume a well-established velocity profile. To achieve this, a sufficient length of straight pipe upstream of the orifice meter is essential. The required straight length of pipe depends on the type of fitting or valve, but can be as much as 40 pipe diameters. For orifice meter installations within a process, attaining this distance is not always easy. A much shorter straight length (2 or 4 pipe diameters) downstream of the orifice meter is also recommended.

For certain types of fittings, the installation of straightening vanes (usually a bundle of thin-walled tubes mounted within the pipe) reduces the required length of straight pipe upstream of the orifice meter.

Integral Orifice Assembly

Most manufacturers of pressure transmitters also supply an *integral orifice assembly*, which attaches directly to the pressure transmitter, thereby effectively creating an orifice flow transmitter. Figure 5.11 illustrates the major components of an integral orifice assembly. The key points are as follows:

- The connecting piping from the orifice meter assembly to the pressure transmitter is eliminated. This reduces installation costs and eliminates sources of leaks.
- Integral orifices are available only for small line sizes, typically 0.5 to 1.5 in. But in such small lines, the integral orifice offers distinct advantages.

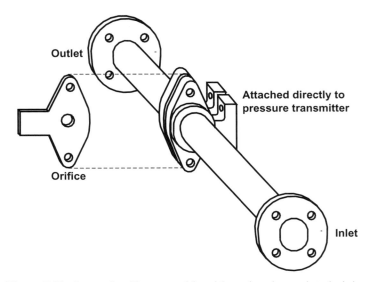

Figure 5.11. Integral orifice assembly with optional associated piping.

The integral orifice provides the precise alignment necessary for small lines.

- Errors from pipe roughness are more significant in small line sizes. Most suppliers of integral orifices can also supply associated piping that addresses these issues. The associated piping illustrated in Figure 5.11 has flanged ends, but normal pipe threads and welded ends are also available.
- The supplier must provide the coefficients that relate flow to pressure drop. Typical accuracy is ±1% of upper-range value.
- Different orifice sizes cover different flow measurement ranges. Integral orifices are capable of measuring very low flows. Upper-range values of 2.5 gph or 24 scfh are possible, with turndown ratios of about 5:1.
- Where functions such as calibration of the pressure transmitter are required without disrupting process operations, a three-valve manifold can be inserted between the pressure transmitter and the integral orifice.
- Issues sometimes arise when the pressure transmitter is in close proximity to the fluid. For example, high fluid temperatures could be a problem. Remote connectors can be inserted between the pressure transmitter and the integral orifice assembly to address such issues.

5.7. HEAD METERS

The following flow meters are all head meters:

- Orifice meter.
- Venturi meter.
- Flow tube.
- Flow nozzle.
- Elbow taps.
- Pitot tube.
- Averaging pitot tube.

The generic equation for all head meters is the same as that for an orifice meter:

$$Q = C_o \left[\frac{\Delta P}{G} \right]^{1/2}$$

where Q = volumetric flow; ΔP = pressure drop at pressure tap locations; G = fluid specific gravity; C_o = orifice coefficient.

A differential pressure transmitter is required to sense the difference in pressure ΔP between a high-pressure tap and a low-pressure tap. The volumetric flow Q is related to the square root of this differential pressure. The value of the coefficient C_o depends on the nature of the head meter. The effect of the specific gravity G on the volumetric flow is the same as discussed for the orifice meter. For gas flow measurements, temperature and pressure compensation are also the same as for the orifice meter.

Venturi Meter

An orifice plate introduces considerable turbulence, and the associated loss in energy is reflected in the permanent pressure loss. As illustrated in Figure 5.12, the venturi meter consists of a *throat* of a smaller diameter than the pipe, with conical sections for entrance and exit to reduce the turbulence. Compared to the orifice meter, the permanent pressure loss is less. But whereas the orifice meter can be inserted at a flange in the piping, the venturi meter occupies a much longer pipe section. The most common use of venturi meters is for applications with large flow rates in which the lower permanent pressure loss offsets such disadvantages.

The equations that describe the venturi meter are similar to those for the orifice meter. The design parameters for a venturi meter are stipulated in ISO 5167-4[4]; PC-based programs are available for designing venturi meters. The physical geometry of a venturi meter is as follows:

- Entrance cone angle α_1 is $21° \pm 2°$.
- Exit cone angle α_2 is between $5°$ and $15°$.
- Throat length is equal to the throat diameter.
- The high-pressure tap is 0.25 to $0.5 D$ upstream of the entrance conical section.
- The low-pressure tap is at the center of the throat.

Flow Tube

Most flow tubes resemble a venturi meter, but the supplier has modified the design to achieve a greater pressure recovery. The design of a flow tube is

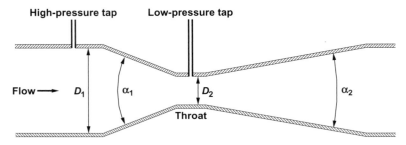

Figure 5.12. Venturi meter.

generally proprietary to the manufacturer of the flow tube. As illustrated in Figure 5.13, the modifications are primarily to the entrance cone; the objective is to reduce the turbulence created when the flow area is reduced. The remainder of the flow tube (expanding cone, pressure taps, etc.) is usually very similar to a venturi meter.

The main incentive for installing a flow tube is energy savings. In large line sizes, reductions in the permanent pressure loss translate into reductions in the energy required at the pump or blower.

The accuracy of a flow tube is comparable to the accuracy of a venturi meter, which is about ±0.5% of the upper-range value. The value for the coefficient C_o is very close to the value for a venturi meter, but because of the proprietary design of the flow tube, the manufacturer must either provide the value of C_o or explain how it is to be calculated.

Flow Nozzle

As illustrated in Figure 5.14, a flow nozzle is basically an aerodynamically improved version of an orifice. However, the permanent pressure loss of a flow

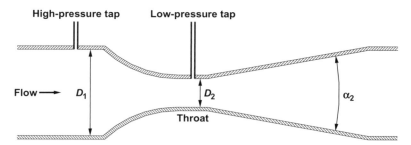

Figure 5.13. Flow tube.

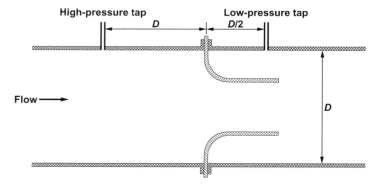

Figure 5.14. Flow nozzle.

nozzle is only slightly less than for an orifice. Flow nozzles are installed in the following situations:

Solids present in the fluid. The smooth entrance section avoids the buildups of solids that would occur in an orifice meter.

Abrasive applications. The accuracy of a flow nozzle is less affected by wear on the entrance section.

High velocities. A flow nozzle will pass more fluid than an orifice of the same diameter.

Some flow nozzles are installed within a smooth bore section of the pipe, either welded in place or held in place via a retaining ring. Flow nozzles can also be installed at a flange, usually with the low-pressure tap incorporated into the nozzle.

The accuracy of a flow nozzle is about ±1% of upper-range value, which is comparable to that of an orifice meter. But like the orifice meter, straight lengths of pipe are recommended upstream and downstream; the lengths depend on the β-ratio (orifice diameter/pipe diameter) and the type of fittings upstream. The upstream length may be as much as 70 pipe diameters (straightening vanes reduce this value); the downstream length may be as much as 8 pipe diameters.

Elbow Taps

As illustrated in Figure 5.15, the low-pressure tap is on the inside of the elbow, and the high-pressure tap is on the outside of the elbow. The differential

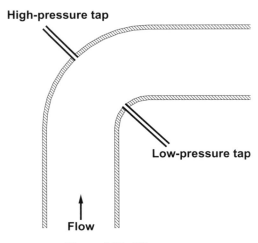

Figure 5.15. Elbow taps.

pressure is the result of the centrifugal forces, which are greater on the outside than on the inside.

The main advantage is that elbow taps can be installed with essentially no changes to the process piping, provided that there is an existing elbow. There is no additional pressure loss beyond that of the piping itself.

The down side to elbow taps is that the accuracy is poor, the usual statements being between ±5 and ±10% of the upper-range value. Custom calibration of the elbow tap installation can improve this somewhat.

Pitot Tube

The pitot tube is an instrument that measures the velocity of the fluid at a specific location within a pipe or duct. As illustrated in Figure 5.16, the pitot tube senses two pressures:

Static pressure. This is the pressure sensed by a pressure tap inserted into a surface that is parallel with the direction of flow. In the illustration, the static pressure is sensed by openings on the side of the pitot tube assembly.

Stagnation pressure. This is the pressure that results when a flowing fluid is brought to rest. The center opening of the pitot tube assembly faces directly into the flowing stream and thus senses the stagnation pressure.

The difference between the stagnation pressure (the high pressure) and the static pressure (the low pressure) is the velocity head of the flowing fluid at that point in the pipe or duct. Most pitot tube assemblies permit the insertion

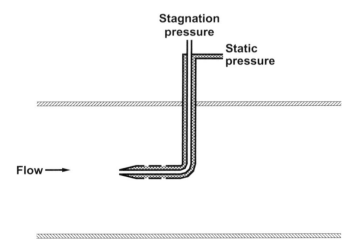

Figure 5.16. Pitot tube.

depth to be adjusted and thus can be used to determine the velocity profile within the pipe or duct. To sense the average velocity, the insertion depth of the pitot tube must be where the point velocity is the same as the average velocity. Figure 5.16 locates the pitot tube at the centerline, but this is not likely to be the appropriate insertion depth. For laminar flow in circular pipes, the average velocity occurs at a distance of $0.33 R$ from the centerline of the pipe.

Averaging Pitot Tube

Figure 5.17 illustrates an annubar, which is one implementation of an averaging pitot tube. Whereas the pitot tube has only one port for sensing the stagnation pressure, the annubar averages the stagnation pressures sensed by multiple ports. The objective is to provide a stagnation pressure that reflects the average velocity of the flowing stream. The key points are these:

- The permanent pressure loss for annubar installations is negligible. The annubar element is inserted into the flow stream but does not significantly reduce the flow area. Some turbulence is introduced into the flowing stream, but it is rarely enough to cause a detectable permanent pressure loss.
- Most annubar elements are designed so that they can be withdrawn without interrupting process operations.
- The annubar senses volumetric flow with an accuracy of about ±1% of upper-range value.
- Unlike the setup in Figure 5.17, the static pressure is sensed via ports incorporated into the annubar elements. Ports on the leading face

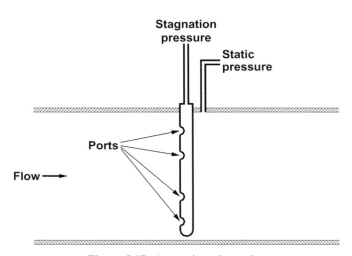

Figure 5.17. Averaging pitot tube.

sense the stagnation pressure; ports on the trailing face sense the static pressure.

- A common application of annubars is to measure the flow in large gas ducts. The duct may be either circular or rectangular.
- The port arrangements on the annubar element are proprietary designs. For most installations, the manufacturer should be consulted regarding the installation of the annubar.

5.8. CORIOLIS METERS

Angular momentum is one potential basis for a flow meter to directly sense mass flow. Unfortunately, such a meter is too mechanical to be practical in an industrial environment. However, angular momentum is a good starting point for understanding the principles on which the coriolis meter relies.

Angular Momentum

To understand this approach, consider the water sprinkler shown in Figure 5.18. The common water sprinkler maintains rotation via a bent arm near the tip. The water jets out at an angle, thus providing the force that causes the sprinkler to rotate.

This is not the case for the water sprinkler in Figure 5.18. The water jets straight out. Rotation at the desired speed will be achieved by applying a torque to the sprinkler. How much torque is required? Let's neglect friction and consider the length R of the sprinkler arm to be fixed. The torque then depends on two factors:

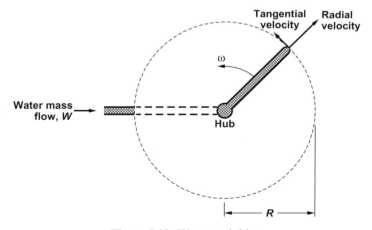

Figure 5.18. Water sprinkler.

- The rotational speed ω.
- The mass flow W of the water.

Velocity of the Water   Consider the velocity of the water as it leaves the tip of the sprinkler. There are two components:

Radial component. This is the velocity of the water relative to the hub of the water sprinkler. This velocity depends on the water supply pressure but otherwise is of no concern herein.

Tangential component. The velocity in the tangential direction increases from zero at the hub to $2\pi R\omega$ at the tip. This velocity depends on the rotational speed.

As shown in Figure 5.19, the tangential velocity increases linearly from zero to $2\pi R\omega$. This requires the application of a constant force of $2\pi\omega W$ per unit of sprinkler arm length. The torque required to apply such a force is $\pi R^2\omega W$. For a given rotational speed ω, the torque required to rotate the sprinkler is directly proportional to the mass flow W of the water.

Coriolis Force   As the water flows from the hub of the sprinkler to the tip, a force is required to accelerate the water from zero to the tip velocity. This requires that a force be applied by the sprinkler arm to the water molecules (Fig. 5.20).

This force, known as the coriolis force, also appears in the following situations:

Merry-go-round. Suppose you walk from the center of a merry-go-round to the outer edge. You are accelerated from a velocity of zero to the

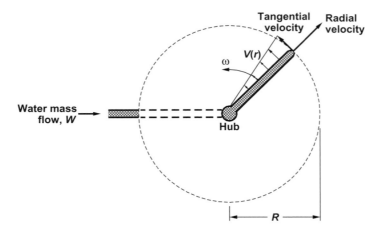

Figure 5.19. Tangential velocity for the water sprinkler.

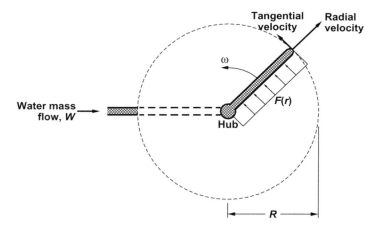

Figure 5.20. Coriolis force exerted by the sprinkler arm on the water.

velocity at the edge. You must lean so that a force is applied in the direction of rotation. If you do not lean, you will fall in a direction opposite to the direction of rotation.

Surface of the earth. Suppose you walk from the north pole to the equator. You are accelerated from a velocity of zero to the surface velocity at the equator (slightly over 1500 ft/sec). This is a substantial increase in velocity, but it occurs over a distance of more than 6000 mi. The required force is so small that you do not notice. However, it is reflected in phenomena such as the direction of the swirl as liquid drains from a vessel.

Deformation   To accelerate the water from zero to the tip velocity, the sprinkler arm must be applying the necessary force to the water. This results in a deformation of the sprinkler arm (Fig. 5.21). This deformation depends on the following:

- Structural characteristics of the sprinkler arm such as the length of the sprinkler arm, diameter of the tube, thickness of the tube, and modulus of elasticity of the tube metal. For a given sprinkler arm, these structural characteristics are fixed, with one exception: The modulus of elasticity of the tube metal depends on tube temperature.
- Mass flow W of the water.

For a given sprinkler arm, the deformation depends on the mass flow W of the water and the tube temperature. The mass flow W of the water can be inferred by sensing the tube deformation and compensating for changes in tube metal temperature.

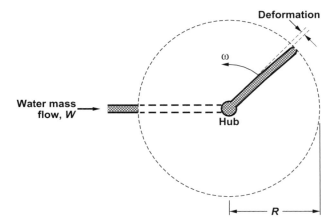

Figure 5.21. Deformation of the sprinkler arm.

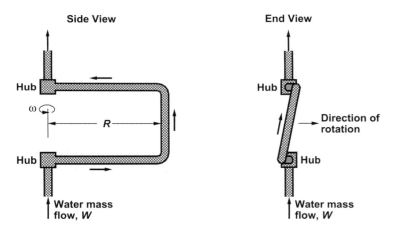

Figure 5.22. Rotating U-tube.

Rotating U-Tube

Sensing the deformation is the first step in developing a practical implementation of a mass flow meter. The next step is to devise a flow-through configuration. This is accomplished by the U-tube configuration illustrated in Figure 5.22. The water enters at the bottom, is accelerated to the tip velocity in the lower arm, and is then decelerated to zero in the upper arm. In this configuration, the deformations are as follows:

Lower arm. The water is accelerating from zero to the tip velocity. The lower arm must be applying the necessary force to the water. This leads to a deformation in the lower arm that is opposite to the direction of the rotation.

Upper arm. The water is decelerating from the tip velocity to zero. The water is now applying a force to the upper arm. This leads to a deformation in the upper arm that is in the direction of the rotation.

The deformations in the lower and upper arms should be identical, but in opposite directions. From an end view of the U-tube, the upper arm is leading the lower arm with respect to the direction of rotation. That is, the U-tube is twisted by an amount that increases with the mass flow of the water.

Vibrating U-Tube

Rotating the U-tube raises a number of rather difficult design issues, such as maintaining leakproof seals at the hubs. Instead of rotating the U-tube, suppose we vibrate the U-tube at a known frequency. This is equivalent to a rotating arrangement in which the direction of rotation reverses twice each cycle.

In Figure 5.23, the U-tube is mounted vertically, with the flow entering at the bottom and exiting at the top. The twisting of the U-tube reverses when the direction of motion is reversed:

Lower leg of the U-tube. The fluid is being accelerated in the direction of motion. This causes the lower leg to deform in a direction opposite to the direction of motion. Basically, this leg always lags with respect to the direction of motion.

Upper leg of the U-tube. The fluid is being decelerated. This causes the upper leg to deform in the same direction as the direction of motion. Basically, this leg always leads with respect to the direction of motion.

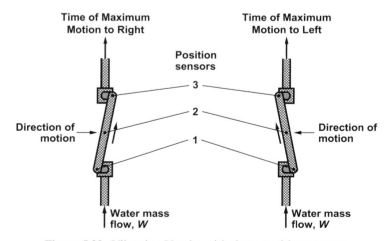

Figure 5.23. Vibrating U-tube with three position sensors.

Figure 5.23 also illustrates three position sensors attached to the U-tube:

Position sensor 1. Mounted at the top of the lower leg of the U-tube.

Position sensor 2. Mounted at the center point between the legs of the U-tube.

Position sensor 3. Mounted at the top of the upper leg of the U-tube.

The vibrator for the U-tube is attached at the location of position sensor 2. Therefore, this sensor indicates the results of the excitation being applied to the U-tube. We have included this sensor only for purposes of reference. In commercial products, this position sensor is not required.

Effect of Flow

The vibrator will essentially cause the position at sensor 2 to vary in a sinusoidal fashion. The amplitude of the sinusoid depends on the energy imparted by the vibrator. The frequency of the sinusoid depends on the structural characteristics of the U-tube and its weight. Commercial mass flow meters can determine the density of the fluid from the frequency, but let's defer this to a later discussion.

When there is no flow, there is no deflection of the U-tube. As shown in Figure 5.24, the positions at sensors 1 and 3 are the same as for sensor 2—that is, all three are identical sinusoids. Specifically, there is no phase shift between any of the sinusoids.

When fluid is flowing through the U-tube, the signals from the position sensors are still sinusoids. But due to the coriolis force in the lower leg of the

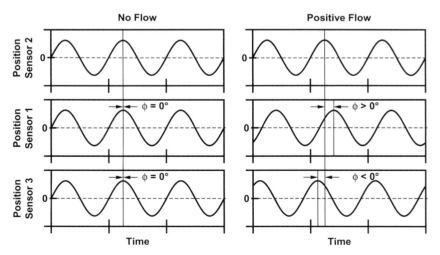

Figure 5.24. Signals from position sensors for vibrating U-tube.

U-tube, the sinusoid from position sensor 1 lags the sinusoid from position sensor 2. The sinusoid from position sensor 3 leads the sinusoid from position sensor 2 by exactly the same amount. This results in a phase shift between the various sinusoids. Commercial mass flow meters use only position sensors 1 and 3. The phase shift between the sinusoids from these position sensors increases as the mass flow increases. The phase shift varies linearly with the mass flow. Basically, the phase shift is sensed and then multiplied by a coefficient to obtain the mass flow.

Micro Motion

The coriolis mass flow meter technology was pioneered by Micro Motion, which is now part of the Emerson organization. The term *coriolis meter* is the generic term that reflects the basic principle on which the meter relies. However, the term *Micro Motion meter* continues to be commonly used, although several manufacturers now produce coriolis meters.

The coriolis meter revolutionized practices within certain industries by providing flow measurements where none previously existed. For example, until the coriolis meter appeared, there was no technology for measuring the flow of an emulsion whose continuous phase is an organic. Such emulsions are encountered in processes for producing adhesives, coatings, paints, etc. Before the coriolis meter, the emulsion flow rate had to be inferred from the change in the level or weight of the vessel supplying the emulsion.

Although the coriolis meter is marketed primarily for measuring the mass flow, the meter can also measure the fluid density. Occasionally, a coriolis meter will be installed where only the fluid density is of interest.

Tube Shapes

Most of the currently installed coriolis meters are classified as bent-tube coriolis meters. The U-tube is a bent tube, but other shapes can and have been used within coriolis meters. The consequences of using bent-tubes include the following

- Pressure drop within the coriolis meter is significant, especially at high velocities. In liquid service, one must make sure that the pressure at the meter discharge is safely above the fluid vapor pressure.
- Bent-tube coriolis meters tend to be bulky pieces of equipment.
- Should cleaning become necessary, the bends in the tube become an obstacle to mechanical cleaning.
- Although the U-tube is the simplest to explain, other bent-tube configurations give a higher sensitivity to flow.

A coriolis meter containing a straight tube was introduced by Krohne in 1994. The pressure drop is less and cleaning is far easier. The sensitivity of the deformations to flow are lower, but this technology is capable of providing the accuracies required in industrial flow meter applications.

Most coriolis meters are delivered with an arrow indicating the direction of flow. This is essentially the direction of positive flow. The coriolis principle works with the flow in either direction, and most microprocessor-based transmitters can be configured to handle reversible-flow applications.

Sensitivity to Vibrations

Any measurement device that relies on vibrations is potentially susceptible to vibrations originating within the process. The early versions of the coriolis meter contained a single tube that was vibrated relative to its enclosure. These units proved to be susceptible to vibrations from the process. This was addressed by securing the meter to the most stable structural member available (usually a large block of concrete). Unfortunately, this practice has a side effect—namely, that the piping connected to the securely mounted coriolis meter produces mechanical stresses on the meter.

The next versions of the coriolis meter contained two tubes that were vibrated relative to each other. This reduced the sensitivity to process vibrations to a very acceptable level (process vibrations severe enough to affect the coriolis meter are usually undesirable for other reasons). Securely mounting these coriolis meters is not recommended. The coriolis meter is a rather bulky piece of equipment that requires some support, but not to an excessive degree. Anything that would cause the meter to be subjected to a mechanical stress must be avoided. Of course, the connecting piping is certainly a potential source of such stresses.

The current direction is to return to single-tube designs. But instead of vibrating the tube relative to the meter enclosure, the tube is vibrated relative to a counterbalanced weight. This also reduces the sensitivity to external vibrations to an acceptable level for industrial applications.

Response Time

The coriolis meter responds very rapidly to changes in the fluid flow. The response is much faster than required for most applications. In fact, the coriolis meter responds so rapidly that its output will reflect any fluctuations in the flow.

One situation in which fluctuations will exist is when the coriolis meter is installed on the discharge of a positive displacement pump. The coriolis meter responds rapidly enough to sense the pulsating flow. The result appears to be a very noisy flow measurement. However, one must resist the temptation to apply smoothing.

The effective solution is to install a pulsation damper on the discharge of the positive displacement pump (Fig. 5.25). A pulsation damper contains a diaphragm with the process liquid on one side and a compressed gas on the other. The gas pressure is usually about 80% of the process operating pressure.

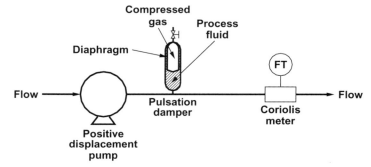

Figure 5.25. Pulsation damper on discharge of positive displacement pump.

The pulsation damper absorbs the pulsations generated by the pump, which gives a smooth flow through the coriolis meter.

Effect of Gas Pockets

Coriolis meters can measure the mass flow of either liquids or gases. Coriolis meters can also measure the mass flow of two-phase mixtures, either liquid–liquid (such as emulsions) or liquid–solid. But when one of the phases is a compressible fluid, problems arise—that is, when applied to measuring the mass flow of a liquid, there must be no gas within the liquid. When applied to measuring the mass flow of a gas, there must be no entrained liquid in the gas.

The presence of a gas within a liquid stream can arise for several reasons:

- Gas is present within the liquid entering the meter. For example, excessive agitation will entrain air into an emulsion.
- The pressure drop within the coriolis meter may drop the pressure below the vapor pressure of the fluid, resulting in flashing.
- Gas pockets may develop within the coriolis meter (dissolved gases are released as the pressure drops within the meter).

The presence of gas within the tubes of a coriolis meter will cause the liquid mass flow measurement to be in error. When designing an installation, always be prepared to address the possibility of trapped gas within the meter (if any problem arises, this is invariably cited as the possible cause). One way to avoid gas pockets is to mount the meter in a vertical line with the flow entering at the bottom (Fig. 5.26).

Low-Density Cutoff

In addition to sensing the mass flow, a coriolis meter also senses the density of the fluid within the tubes. If the fluid density indicated by the meter is less

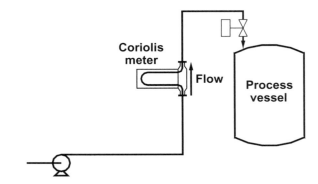

Figure 5.26. Avoiding gas pockets by mounting meter vertically.

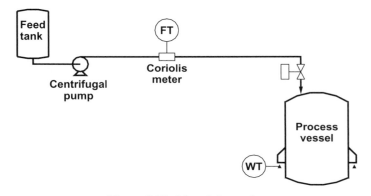

Figure 5.27. Material transfer.

than the expected density of the fluid, some gas must be present within the meter.

In some applications, problems are avoided by using a low-density cutoff. This logic is simple: If the density is less than a specified value, then set the mass flow to zero. In the example given in Figure 5.27, a material is being pumped from a feed vessel to a process vessel. A coriolis meter senses the mass flow. Air can get into the meter due to the following:

- When no transfer is in progress, the pump is stopped. Depending on the configuration of the piping, the fluid can *bleed back*, introducing air into the coriolis meter.
- In some batch applications, a mixture is prepared in the feed vessel, all of which must be pumped into the process vessel. At the end of the transfer, it is possible that air is introduced into the coriolis meter.

A low-density cutoff will reduce (if not eliminate) the error in situations such as that shown in the figure. But consider pumping an emulsion. If the emulsion is agitated excessively (an abnormal situation), air will be entrained into the emulsion. A low density suggests that the measurement is inaccurate, but there is no way to obtain an accurate value for the mass flow.

Advantages

The advantages of a coriolis meter include the following:

- Senses mass flow, not volumetric flow.
- Not affected by fluid properties.
- Insensitive to Reynolds number.
- No upstream or downstream straight pipe required.
- Very accurate ($\pm 0.1\%$ of upper-range value).
- High turndown ratio (10:1 or better).
- Very stable.
- No sources of leaks within the meter.

The major advantage of the coriolis meter is that it very accurately senses mass flow. There is no need to compensate the measured value for changes in fluid density. Nor is the meter affected by any other fluid properties (viscosity, temperature, pressure, etc.).

The tube can be manufactured from a variety of materials, so a suitable material can be found for most corrosive applications. It is also possible to use lined tubes. Compensation is provided for changes in the temperature of the tube metal. Versions for high-pressure (4000 psig) applications are available. For high-temperature applications, the transmitter can be remotely mounted.

The coriolis meter is very stable. Small errors in the flow measurement are unusual; either the meter is dead on or grossly in error. Users can adjust certain parameters within the transmitter, but such adjustments should be infrequent. Only the manufacturer can repair a coriolis meter, but repairs should also be infrequent.

Disadvantages

The disadvantages of a coriolis meter include the following:

- Expensive.
- High pressure drop.
- Bulky item of equipment.
- Affected by gas pockets (liquid applications).

- Affected by mechanical stresses.
- Large sizes (>2 in) not widely available.

The purchase price is usually the main obstacle to installing a coriolis meter. The market is currently quite competitive, which is having the usual effect on prices. The low maintenance requirements for a coriolis meter make the life-cycle costs somewhat more favorable. However, most coriolis meters are installed with the expectations of improvements in process operations.

Coriolis meters are the meter of preference in industries such as specialty chemicals, pharmaceuticals, and other industries in which the line sizes tend to be small. Meter cost and bulkiness increase rapidly with line size. These seem to be manageable for sizes up to about 2 in. The largest size currently available is 12 in.

The presence of gas pockets within the meter is the most common source of measurement errors for a coriolis meter in liquid service. When investigating problems with a coriolis meter, one of the first actions by service personnel is to check the density indicated by the meter. A value lower than expected suggests that gas is somehow getting into the tubes of the meter.

Material Transfer

In the batch process shown in Figure 5.27, a specified amount of the material is to be transferred from the feed tank to the process vessel. The transfer can be based on either

- The weight transmitter (load cell) on the process vessel.
- The coriolis flow meter.

In the example in the figure, the destination vessel for the material transfer is equipped with a load cell. The issues are the same when the source vessel is equipped with a load cell, except that the source vessel is less likely to be agitated, to have a jacket, or to have other factors that degrade the performance of the load cell.

In batch applications, charging accuracy is usually extremely important. Therefore, the raw material transfer should be based on the measurement that is considered to be the most accurate.

Load Cells

Load cells are generally viewed as the most accurate measurement available. In solids-blending applications, the solid feeds are first discharged onto a pan that is suspended via a load cell. The specified amount is discharged into the pan, and the material is subsequently dumped into a blender. The weight on the load cell is entirely vertical, which is the ideal situation for the load cell.

But when weighing fluids, numerous complications arise. The load cell can be installed on either the destination vessel or the source vessel. For the vessel on which the load cell is installed, only one transfer in or out can be occurring at a given time, which imposes some constraints on the operation of a batch process.

How to properly install a load cell is well known. But in most process load cell installations, compromises are made, especially when the process material is

- Hazardous.
- At a high temperature.
- Under high pressure.

Such characteristics cause objections to be raised to items like flexible connectors. Even when every compromise is made for good reasons, the consequence is the same: Every compromise degrades the performance of the load cell.

Horizontal Stresses

Fluids flow through rigid pipes, which can impart horizontal stresses onto the load cell. Horizontal stresses on a vessel lead to errors in the vessel weight as determined via a load cell. Current load cell technology can very accurately measure the stress on a supporting member. Unfortunately, the load cell cannot discriminate between vertical stresses and horizontal stresses. The vessel design and load cell installation must be such that weight leads to vertical stresses only. Usually this is done quite well, especially for vessels whose original design contemplated the installation of a load cell.

Process vessels always have several piping connections (for feeding materials, for pressurizing or evacuating, for discharging, etc.). Avoiding stresses from these connections is essential, hence the need for flexible connectors. But to those responsible for personnel safety, flexible connectors (even metal-clad flexible connectors) are a cause for concern. One alternative is to obtain the desired flexibility through long lengths of pipe. Unfortunately, long lengths lead to stresses resulting from thermal expansion and contraction.

Total Weight

A load cell sees total weight. Normal practice is to zero, or tare, the load cell with the vessel empty. Any subsequent change in the vessel weight is reported as a change in the weight of the contents.

Some vessels are equipped with a coil or jacket. Any change in the weight of the fluid within the coil or jacket will be reflected in the weight indicated by the load cell. This is especially large when the fluid may change (water vs. glycol vs. steam). But even if always full of water, temperature affects water density and thus the mass of the water within the coil or jacket.

In a large vessel, simply pulling a vacuum will affect vessel weight. At atmospheric pressure and 60 °F, a 1000-gal vessel contains about 10 lb air. Therefore, pulling a vacuum on the vessel would effectively reduce the weight of the vessel by 10 lb. The effect can be even larger for pressurized vessels.

When the vessel is equipped with an agitator, some noise is invariably imparted to the load cell. Usually this leads to filtering or smoothing, either within the load cell electronics or within the controls. For a reading taken when the vessel weight is constant, filtering or smoothing does not affect the value. But when on-the-fly readings are taken while a material transfer is in progress, the effect of filtering or smoothing is to *lag* the indicated weight. When transferring material into the vessel, the actual weight is greater than the indicated weight; when transferring material out of the vessel, the actual weight is less than the indicated weight.

Test

The installed accuracy of a coriolis meter will be very close to the accuracy stated by the manufacturer, typically ±0.1% of upper-range value or better. The installed accuracy of a load cell depends on so many factors that it can be resolved only via a test. The load cell is usually calibrated by placing weights on the vessel supporting members and comparing the indicated weight. However, the following test reflects the manner in which the load cell readings are used during production operations:

1. Start with the vessel partially filled, agitator running, heating/cooling active, etc. Record the weight indicated by the load cell. Then stop all agitation, heating/cooling, etc. Wait until the load cell reading is constant, and record the value.
2. Start agitator, heating/cooling, etc. Then transfer a specified amount of material into the vessel based on the coriolis meter.
3. Record the weight indicated by the load cell. Then stop all agitation, heating/cooling, etc. Wait until the load cell reading is constant, and record the weight.

The weights taken with agitator running, heating/cooling active, etc. are the on-the-fly weight readings. These rarely agree with the coriolis meter. But often the equilibrium weight readings agree quite well with the coriolis meter. If so, relying on the coriolis meter will produce the most accurate material transfers.

5.9. MAGNETIC FLOW METER

The basis for the magnetic flow meter is Farraday's law: If a conductor moves through a magnetic field, an electrical potential is generated that is proportional to

- The velocity of the conductor.
- The strength of the magnetic field.

From the perspective of magnetic flow meters, of equal importance is what is not included in this list: the conductivity of the conductor. If a copper conductor and an aluminum conductor move at the same velocity through the same magnetic field, the same electrical potential is generated.

DC generators also rely on Farraday's law. The armature wiring in such generators is normally copper. Copper has a high conductivity (low resistance). The power loss and the heat generation within the armature is I^2R, where R is the resistance of the armature wiring and I is the current flow. Less power is lost and less heat is generated when copper is used. However, if the armature wiring were aluminum, the only difference in the output voltage would be due to the IR voltage drop within the armature.

Components

As Figure 5.28 suggests, a magnetic flow meter basically consists of three components:

- A liner that is not ferromagnetic. This can be nonferromagnetic metals (such as stainless steel), plastic, ceramic, etc.
- Electromagnets to generate the magnetic field.
- Electrodes that make contact with the fluid. If the tube is conducting, these must be insulated from the tube.

The transmitter senses the voltage that is generated within the magnetic flow meter and generates an output signal that is linear with respect to the

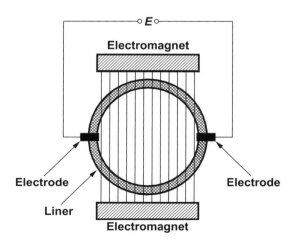

Figure 5.28. Magnetic flow meter.

volumetric flow. All of these components are enclosed in a housing consistent with the process pressure rating. The transmitter can be mounted locally or remotely.

The fluid flowing through the pipe can be viewed as a conductor moving through a magnetic field. The electric potential that is generated is proportional to the average velocity V of the fluid. The volumetric flow Q is the product of the average velocity V of the fluid and the cross-sectional area A of the tube. The magnetic flow meter is inherently linear—the volumetric flow Q is proportional to the voltage across the electrodes.

Conductivity

All specifications for magnetic flow meters stipulate that the fluid conductivity must be greater than some value. As stated previously, the electrical potential generated within the magnetic flow meter does not depend on the conductivity. So why the specification on conductivity?

Suppose the conductivity of the conductor is approaching zero. This has no effect on the electric potential that is generated. The limiting condition is an insulator. But if an insulator how would you measure whatever potential, if any, that is generated? All voltage-sensing devices must flow a small current around a circuit. The impedance (resistance) of the voltage-measuring device must be much greater than the impedance of the remainder of the circuit. As the conductivity of the fluid decreases, the impedance of this part of the circuit increases. At some point, this impedance becomes a significant part of the total impedance of the circuit, resulting in errors in the voltage measurement.

The requirement on minimum conductivity depends on the voltage-sensing mechanism within the transmitter, specifically, its impedance. To decrease the minimum required fluid conductivity, the manufacturer must increase the impedance of the voltage-sensing mechanism. Most manufacturers make this impedance as high as possible, given the current state of the art.

Electromagnets

Most commercial magnetic flow meters use a pulsed excitation (several times per second) of the electromagnets instead of a continuous excitation. The main benefit of this is to reduce the power required by the electromagnets. One is not concerned about the cost of this power. There are three objectives:

- Reduce the power requirements so that a two-wire transmitter can be used.
- Reduce the heat generation within the electromagnets.
- Reverse the polarity between pulses to reduce the effect of noise.

The heat generated within the electromagnets is dissipated into the flowing fluid. The quantity of heat is so small that it has no detectable effect on fluid temperature. However, as the temperature of the fluid increases, the heat generated within the electromagnets is not dissipated, leading to excess temperatures within the electromagnets. This issue leads to a specification on the maximum fluid temperature, which is typically 400° to perhaps 500 °F.

Velocity Profile

The magnetic flow meter senses the average velocity of the fluid between the two electrodes. This makes the meter insensitive to

Reynolds number. The meter works in both laminar and turbulent flow. The typical turndown ratio of 50:1 permits the meter to be installed in applications in which the flow is high at times but low at other times.

Fluid properties. This includes viscosity and density. The magnetic flow meter is a volumetric meter. Changes in the density have no effect on the volumetric flow sensed by the meter. An accurate density is required to convert the volumetric flow to a mass flow.

What about straight lengths of pipe upstream and downstream? You will find industrial installations of magnetic flow meters with elbows immediately upstream and/or downstream. However, this is not generally recommended. The meter senses the average velocity of the fluid between the two electrodes. Therefore, distortions in the velocity profile would have some effect on the meter. Although the degree is somewhat in dispute, these tend to be small. The accuracy of the magnetic flow meter is typically stated as ±0.5% of upper-range value. Minor distortions in the velocity profile would not be noticed.

Advantages

The advantages of a magnetic flow meter include the following:

- No obstruction to flow.
- No pressure drop; no flashing.
- High turndown ratio (50:1).
- Very stable (low drift).
- Used for corrosive fluids.
- Applicable to reversible flows.
- Not damaged by excess fluid velocities.
- Can be installed horizontally or vertically.
- No sources for fluid leaks.
- Rugged meter with low maintenance.

If one looks through a magnetic flow meter, one basically sees a smooth pipe. The electrodes can be seen, but protrude little if at all into the flowing stream, resulting in the following:

- Except when the meter size is smaller than the line size, there is no pressure drop and no flashing.
- There is nothing to cause problems with slurries. One application is to measure stock flows (a fiber–water slurry) in the paper industry.
- Two-phase liquid–liquid mixtures, including emulsions, can be metered, provided the continuous phase has the necessary conductivity.

The liner can be manufactured from a wide variety of materials, so a magnetic flow meter can usually be applied in corrosive situations. The magnetic flow meter is routinely used to meter sulfuric acid and caustic. These easily meet the minimum conductivity requirements. The conductivity is a function of acid or caustic concentration, but the magnetic flow meter is not affected by changes in fluid conductivity, provided the conductivity remains above the minimum.

Disadvantages

The disadvantages of a magnetic flow meter include the following:

- Requires minimum fluid conductivity.
- Not applicable to gases.
- Volumetric meter.
- Moderate accuracy ($\pm 0.5\%$ or better).
- Electrode coating is possible.
- Price increases rapidly with size.

The most common factor that eliminates the magnetic flow meter from consideration for a given application is the required minimum fluid conductivity. Although distilled water does not have the required conductivity, the conductivity of most industrial aqueous streams is safely above the minimum. Most organics and oils do not have the required conductivity. But most organic acids and some alcohols provide the minimum conductivity. For other substances, the conductivity must be investigated. The specification on minimum conductivity continues to decline, but liquids such as propane are basically nonconducting. No gases have the required conductivity.

The magnetic flow meter is a volumetric meter with moderate accuracy. It is definitely superior to orifice meters, but not as good as a coriolis meter, even when an accurate value for the density is available to convert from volumetric flow to mass flow.

Coating of the electrodes occasionally leads to problems. Maintaining a high velocity through the meter will minimize electrode coating. Just the mention of electrode coating often raises eyebrows due to the severity of this problem for pH electrodes. However, the electrodes in the magnetic flow meter are not electrochemical electrodes, so the problems are far less severe.

5.10. VORTEX-SHEDDING METER

The basic principle behind the vortex-shedding meter is the von Karman effect. Within a vortex-shedding meter, a blunt object is inserted into the pipe (Fig. 5.29). Vortices are formed on the face of the object and are shed, alternating from one side to the other. The rate (frequency) at which the vortices are formed and shed is a function of

- Fluid velocity.
- Dimensions of the pipe and the blunt object.

For a given meter design, the geometry is fixed. Consequently, the vortex-shedding rate is a function of the fluid (gas or liquid) velocity, but with one proviso: the flow region must be turbulent.

A vortex is a localized region of high velocities. The turbulence results in a permanent pressure loss of approximately two velocity heads. A decrease in pressure is also associated with each vortex. To avoid flashing within a vortex-shedding meter, the static pressure should be about 25% above the fluid vapor pressure.

Digital Meter

The vortex-shedding meter is a digital meter. A mechanism is required for detecting the vortices. Either a vortex is or is not present at the location of the

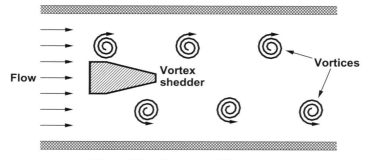

Figure 5.29. Vortex-shedding meter.

detector. Therefore, the output of the vortex-detecting mechanism is a yes–no indication. This is a frequency signal. The mechanism will transition from no to yes some number of times per second, this being the vortex-shedding frequency (number of vortices per second).

To convert the frequency signal to a flow rate requires a calibration constant for the meter. This calibration constant is the volume of fluid per vortex. Therefore, the flow rate is the frequency of vortex shedding times this calibration constant.

The calibration constant is indeed a constant over a fairly wide range of Reynolds numbers. But with the advent of digital transmitters, the conversion of the frequency to flow is enhanced by considering the calibration constant to be a function of the Reynolds number.

Strouhal Number

Dimensionless groups are widely used in fluid dynamics. The Reynolds number is encountered most frequently. For the vortex-shedding resulting from the von Karman effect, the relevant dimensionless group is the Strouhal number N_{St}. For a vortex-shedding meter, the Strouhal number is defined as follows:

$$N_{St} = \frac{fL}{V}$$

where f = vortex-shedding frequency (sec^{-1}); L = width of the blunt face of the vortex shedder (ft or m); V = fluid velocity (ft/sec or m/sec).

The Strouhal number depends only slightly on the Reynolds number. In vortex-shedding meters with analog transmitters, a constant value is used for the Strouhal number. But with digital transmitters, the dependence of the Strouhal number on the Reynolds number can be included.

Detecting the Vortices

Detecting the vortices has proven to be the most difficult aspect of a vortex-shedding meter, with the associated problems giving the early models of vortex-shedding meters a bad reputation. A vortex is a localized region of high velocity and consequently of low pressure. The mechanism for detecting the vortices often determines the temperature limit for the meter. Five of these are

Piezoelectric crystals. Changes in pressure change the stress on the piezo-electric crystal and its frequency of oscillation.

Shuttle ball. With vortices alternately shedding from one side to the other, the resulting pressure changes cause the shuttle ball to oscillate, which is detected magnetically.

Heated thermistor. When a vortex passes over the thermistor, the high velocity cools the thermistor. Even small temperature changes translate into significant changes in the resistance of the thermistor.

Twisting action on the vortex shedder. As vortices are shed in alternate directions, the changes in pressure cause the vortex shedder to wobble or fishtail. Strain gauges can detect the resulting stresses on the support members that hold the vortex shedder in a fixed position.

Ultrasonics. The vortices downstream of the vortex shedder can be detected by ultrasonic transmitters and receivers.

Turndown Ratio

Vortex-shedding meters are capable of high turndown ratios, but with one stipulation: Turbulent flow is a necessary prerequisite for the von Karman effect. That is, in laminar flow, the vortices are not formed, and the vortex-shedding meter will indicate a flow of zero. If one slowly decreases the flow through a vortex-shedding meter, the measured flow will track the actual flow until the transition to laminar flow occurs. At this point, the measured flow abruptly drops to zero.

The minimum line size for a vortex-shedding meter is about 0.5 in; the meter can be supplied in as large a line size as required. The capacity of a given size of a vortex-shedding meter is often stated for water and for air. For each, the minimum and the maximum capacity are stated. The minimum capacity is determined by the transition from turbulent to laminar flow.

Turndown ratios as high as 100:1 have been stated for a vortex-shedding meter. But this is subject to the flow being turbulent over this range. In practice, a transition from turbulent to laminar flow usually occurs before the flow decreases by a factor of 100.

Advantages

The advantages of a vortex-shedding meter include the following:

- Linear digital meter.
- Permanent pressure loss is small.
- Applicable to liquids and gases.
- Cost is competitive with orifice meter.
- No leaks from pressure taps and connecting piping.
- Typical accuracy is ±0.5% of reading.

The vortex-shedding meter is most often considered as an alternative to the orifice meter. On an installed cost basis, the vortex-shedding meter is competitive in price with the orifice meter in all line sizes (from 0.5 in up). In addition, there are no pressure taps and connecting piping for potential sources of leaks.

The permanent pressure loss is relatively small (approximately two velocity heads), which is a definite advantage over the orifice meter in large line sizes for which pumping costs are significant.

The accuracy of a vortex-shedding meter is moderate but is generally stated as a percent of reading. As the flow decreases, the vortex-shedding meter will retain its accuracy, generally up to the onset of laminar flow.

Disadvantages

The disadvantages of a vortex-shedding meter include the following:

- Does not work in laminar flow.
- Upstream and downstream straight pipe required.
- Buildups on the obstruction degrade performance.
- Volumetric meter.

Even though the turndown ratio is potentially as high as 100:1, the lower limit on the measurable flow is usually determined by the onset of laminar flow. Other meters become inaccurate at low flows; the vortex-shedding meter reads zero at low flows. However, the vortex-shedding meter generally retains its accuracy up to the onset of laminar flow.

A well-established velocity profile is required within the meter. This leads to recommendations for straight lengths of pipe upstream and downstream. These recommendations are similar to, but not identical to, those for the orifice meter.

Any buildups on the vortex shedder alter the meter calibration factor—namely, the volume of fluid per vortex. The vortex-shedding meter is potentially applicable to metering slurries, but most slurries would lead to buildups on the vortex shedder. An abrasive slurry is unlikely to buildup on the vortex shedder but would probably change its shape (and thus the meter's calibration factor) through abrasion.

5.11. TRANSIT-TIME ULTRASONIC FLOW METER

Figure 5.30 depicts a wetted-surface ultrasonic flow meter with the electronics mounted so that two ultrasonic beams are generated, one in each direction. The ultrasonic beams are transmitted alternately between the pair of transmitters and receivers.

For a pipe of diameter D, the distance between the physical location of the transmitter and receiver pairs is approximately $D/2$, which means that the angle of incidence θ for the ultrasonic beam is about $30°$ (exactly $D/2$ would give $\theta = 26.6°$).

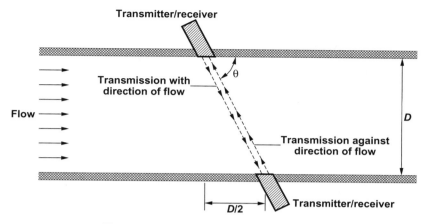

Figure 5.30. Transit-time ultrasonic meter.

The path length X_0 for each ultrasonic beam is given by the following equation:

$$X_0 = \frac{D}{\cos \theta}$$

Under zero flow conditions, the transit time t_0 is the path length divided by the velocity V_S of sound in the fluid:

$$t_0 = \frac{D}{V_s \cos \theta}$$

Transit with Direction of Flow

For the transmission in the direction of fluid flow, the actual path is illustrated in Figure 5.31. This path is a combination of:

Path relative to the flowing fluid. The length of this path is X_1, which is shorter than X_0. The transit time t_1 is the time required for sound to transit this path, which is X_1/V_S. The angle of incidence of this path is less than θ, but for $V \ll V_S$ (fluid velocity much less than sonic velocity), the difference is extremely small.

Fluid velocity. For a transit time of t_1, the contribution of the fluid velocity to the path is Vt_1. The contribution of this to the path of the ultrasonic burst is $Vt_1 \sin \theta$.

The actual path length X_0 is still $D/\cos \theta$. For $V \ll V_S$, the path length X_1 and transit time t_1 are

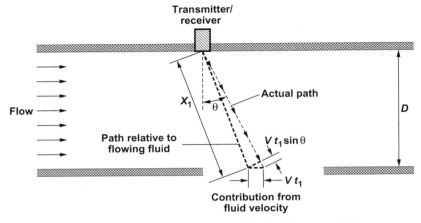

Figure 5.31. Path for transmission with the direction of flow.

$$X_1 = X_0 - Vt_1 \sin\theta$$

$$t_1 = \frac{X_1}{V_S} = \frac{X_0 - Vt_1 \sin\theta}{V_S} = t_0 - \frac{Vt_1 \sin\theta}{V_S}$$

The path length and transit time decrease with the flow.

Transit Against Direction of Flow

By a similar analysis, the path length X_2 and transit time t_2 increase with the flow:

$$X_2 = X_0 + Vt_2 \sin\theta$$

$$t_2 = \frac{X_2}{V_S} = \frac{X_0 + Vt_2 \sin\theta}{V_S} = t_0 + \frac{Vt_2 \sin\theta}{V_S}$$

Meter Equation

The transit-time ultrasonic flow meter senses the following transit times:

t_1, transit time with direction of flow.
t_2, transit time against direction of flow (since $t_2 > t_1$, define $\Delta t = t_2 - t_1$).
t_0, transit time with no flow (the average of t_1 and t_2: $t_0 = (t_1 + t_2)/2$).

The difference Δt in transit times depends on the flow velocity V:

$$\Delta t = t_2 - t_1 = \frac{V(t_1 + t_2)\sin\theta}{V_S} = \frac{2Vt_0\sin\theta}{V_S}$$

This expression is easily solved for the flow velocity V:

$$V = \frac{\Delta t V_S}{2t_0\sin\theta}$$

It is important that the sonic velocity V_S be eliminated from the meter equation. The sonic velocity is not a constant. It is specifically affected by temperature, composition, and density (which for gases is affected by pressure). To obtain the final meter equation, the expression for transit time t_0 is solved for the sonic velocity V_S and then substituted for V_S in the meter equation:

$$V = \frac{\Delta t V_S}{2t_0\sin\theta} = \frac{\Delta t D}{2t_0^2\sin\theta\cos\theta} = \frac{\Delta t D}{t_0^2\sin 2\theta}$$

The final expression for the meter equation relates the mean velocity V to the difference Δt in the transit times, the pipe inside diameter D, the zero-flow transit time t_0 (which is the average of the transit times), and the angle of incidence θ. With this approach, changes in the sonic velocity do not affect the meter.

Multiple Transits

The accuracy and resolution of an ultrasonic meter depend on the length of the path between the pair of transmitters and receivers. Installation of single-transit ultrasonic meters in pipes less than 3 in in diameter should be examined very carefully.

Multiple transits are always in even numbers, which permits the transmitters and receivers to be mounted on the same side of the pipe (rail mechanisms facilitate precise alignments). Figure 5.32 presents configurations for two transits and four transits. The W arrangement is claimed to apply to pipes as small as 0.5 in. Obviously arrangements for even more transits are possible.

However, there are downsides to multiple transits:

- The greater the number of transits, the cleaner the fluid must be. Fortunately, the fluids in some applications (for example, natural gas) are very clean.
- The ultrasonic beam must be reflected one or more times from the pipe walls. Smooth pipes are required; rough pipe walls tend to disperse the ultrasonic beam. Scales and other buildups on the inside of the pipe are definitely bad.

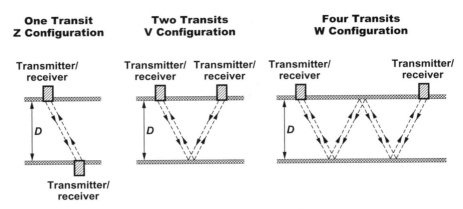

Figure 5.32. Multiple transits.

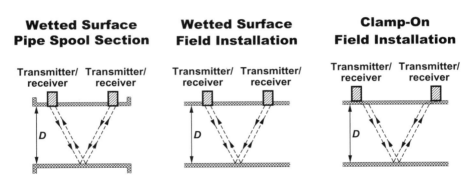

Figure 5.33. Fabrication alternatives.

Fabrication

Figure 5.33 illustrates three options for fabricating an ultrasonic flow meter:

- Wetted-surface transmitters and receivers, installed by the manufacturer into a pipe spool section.
- Wetted-surface transmitters and receivers, installed in the field.
- Transmitters and receivers mounted external to the pipe (usually referred to as clamp-on versions).

The advantage of the pipe spool section is that the manufacturer is responsible for the proper installation and alignment of the transmitters and receivers. In field installations, the user is responsible for properly installing and aligning the transmitters and receivers.

The obvious advantage of the clamp-on versions is that the meter components do not come into contact with the process fluid in any way, thus avoiding issues such as the effect of a corrosive fluid on the transmitters and receivers. But in the clamp-on versions, the ultrasonic beam must travel through the pipe walls. The meter equations must be modified to account for this time (some models sense the thickness of the pipe). But issues arise as to how much of the ultrasonic beam is transmitted through the pipe walls vs. how much of the ultrasonic beam is reflected into the fluid.

Advantages

The advantages of a transit-time ultrasonic flow meter include the following:

- Linear bidirectional volumetric meter.
- Applicable to liquids and gases.
- No permanent pressure loss.
- No flashing or cavitation within meter.
- Cost is almost independent of line size.
- Clamp-on versions are noncontact and leak free.
- Turndown ratios up to 100:1.

The ultrasonic flow meter is a linear volumetric meter that basically senses the average velocity of the flowing stream. The capacity range of the meter is best stated in terms of mean fluid velocity, the typical limits being 0.1 to 100 ft/sec. The meter can sense bidirectional flow of either liquids or gases. Turndown ratios up to 100:1 are stated for some models.

There is no obstruction to flow within an ultrasonic flow meter. There is no permanent pressure loss; there are no regions of low pressure that can lead to flashing or cavitation. Clamp-on versions do not contact the process fluid, so the effect of corrosion and other fluid characteristics on the meter components is not an issue. In toxic applications, clamp-on versions can be serviced with no risk of exposing personnel to the process fluid.

In large pipes, ultrasonic flow meters enjoy a cost advantage. The cost does increase with line size, but only nominally. The cost of a 30-in meter is perhaps twice that of a 3-in meter.

Disadvantages

The disadvantages of a transit-time ultrasonic flow meter include the following:

- Entrained gas bubbles reflect the ultrasonic beam.
- Solid particles scatter the ultrasonic beam.

- Scale and buildups on pipe walls cause problems.
- Upstream and downstream straight pipe recommended.
- Typical accuracy is ±1.0%.
- Temperature limit is 200 °C or less.

The transit-time ultrasonic flow meters depend on being able to transmit sound through the fluid. Entrained gas bubbles tend to reflect sound; solid particles tend to disperse sound. Digital signal processing has improved the ability to precisely detect when the sonic burst arrives at the receiver and to reject false reflections from bubbles. However, transit-time ultrasonic flow meters work best on clean fluids. Scale and buildups on pipe walls also lead to problems, especially in the multitransit versions.

The recommended length of straight pipe upstream ranges from 10 to 50 pipe diameters, depending on type of fitting. Downstream, the typical recommendation is 3 to 5 pipe diameters.

Without custom calibration, the accuracy of an ultrasonic meter is typically in the range of ±1.0% (some state of reading; others of upper-range value). However, accuracy statements as good as ±0.5% and as poor as 3.0% are stated. Wetted-surface versions are generally superior to clamp-on versions. But with custom calibration, ultrasonic meters are now accepted in some custody transfer applications.

5.12. DOPPLER ULTRASONIC FLOW METER

In 1842, Christian Johann Doppler postulated what has become known as the Doppler effect. The frequency of a sound wave depends on the motion of the source (transmitter) and observer (receiver) relative to the propagating medium. In a Doppler flow meter, it is the propagating medium (the fluid) that is in motion; the transmitter and receiver are at fixed positions. A few models of Doppler ultrasonic flow meters mount a transmitter or receiver on a probe that is inserted into the flowing stream. Although rarely acceptable in industrial applications, let's start by examining the principles behind these.

In the upper part of Figure 5.34, a transmitter at a fixed position transmits sound directly upstream into a fluid that is moving at velocity V. The frequency at the transmitter is f_T. But a receiver traveling at the same velocity as the fluid would sense a frequency f_F. Frequency f_F would be higher than frequency f_T, the difference depending on the fluid velocity.

Suppose the sound is reflected by an object that is moving at the same velocity as the fluid. The frequency of the reflected sound would also be f_F (as sensed by a receiver moving at the same velocity as the fluid). But a stationary receiver would sense a frequency f_R. This frequency f_R would be higher than frequency f_F, the difference depending on the fluid velocity.

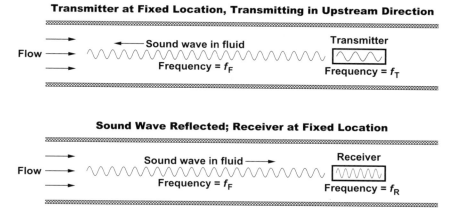

Figure 5.34. Doppler effect.

Frequency Shifts

The first frequency shift occurs between the frequency f_T of the sound generated by the transmitter to the frequency f_F of the sound within the flowing fluid (as sensed by a receiver moving at the same velocity as the fluid). The second frequency shift occurs between the frequency f_F of the sound within the flowing fluid (as sensed by a receiver moving at the same velocity as the fluid) and the frequency f_R sensed by a stationary receiver. These frequency shifts are

$$f_F = \frac{f_T(V_S + V)}{V_S} = f_T + f_T \frac{V}{V_S}$$

$$f_R = \frac{f_F(V_S + V)}{V_S} = f_F + f_F \frac{V}{V_S}$$

The total frequency shift Δf is as follows:

$$\Delta f = f_R - f_T$$
$$= f_F + f_F \frac{V}{V_S} - f_T$$
$$= f_T + f_T \frac{V}{V_S} + \left[f_T + f_T \frac{V}{V_S} \right] \frac{V}{V_S} - f_T$$
$$= 2f_T \frac{V}{V_S} + f_T \left[\frac{V}{V_S} \right]^2$$
$$\cong \frac{2f_T V}{V_S} \qquad \text{for } V_S \gg V$$

Doppler Meter

In most Doppler ultrasonic meters, a single transmitter and receiver unit is mounted on the side of the pipe, either as

Wetted-surface. The transmitter and receiver unit extends through pipe wall, making direct contact with the fluid.

Clamp-on. The transmitter and receiver unit is mounted on the outside of the pipe. This version is the most common.

These transmit sound into the fluid at an angle of incidence θ (Fig. 5.35). Although less common, other variations are occasionally encountered:

Transmitter and receiver mounted separately. The transmitter and receiver can be either on the same side of the pipe or on opposite sides of the pipe. Sound is still transmitted into the fluid at an angle.

Insertion probe. This permits sound to be transmitted directly upstream (as per the previous discussion), but now a probe extends into the flowing fluid.

Frequency Shift

We shall develop the meter equation for a Doppler flow meter that incorporates the transmitter and receiver into the same module. When sound of frequency f_T is transmitted at angle of incidence θ into a fluid flowing at velocity V, the frequency shift Δf of the sound at the receiver is

$$\Delta f = \frac{2 f_T V \sin\theta}{V_S}$$

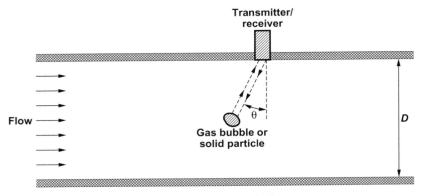

Figure 5.35. Doppler ultrasonic flow meter.

There are two problems in applying this equation directly:

- The velocity of sound V_S in the fluid varies with fluid temperature, fluid composition, and fluid density.
- The angle of incidence θ within the fluid is not the same as the angle of incidence θ_T within the transmitter.

Both of these issues can be addressed by applying Snell's law.

In the wetted-surface transmitter and receiver unit illustrated in Figure 5.36, θ_T is the angle of incidence for the sound directed by the transmitter toward the transmitter–fluid interface, and V_T is the velocity of sound within the transmitter medium. The angle of incidence θ within the fluid is related to the angle of incidence θ_T within the transmitter by Snell's law:

$$\frac{\sin\theta_T}{V_T} = \frac{\sin\theta}{V_S}$$

This relationship also holds for clamp-on versions of transmitters and receivers. Let θ_P be the angle of incidence within the pipe wall and V_P be the sonic velocity within the pipe wall. Snell's law is

$$\frac{\sin\theta_T}{V_T} = \frac{\sin\theta_P}{V_P} = \frac{\sin\theta}{V_S}$$

Substituting Snell's law into the expression for the frequency shift Δf gives the following expression:

$$\Delta f = \frac{2f_T V \sin\theta}{V_S} = \frac{2f_T V \sin\theta_T}{V_T}$$

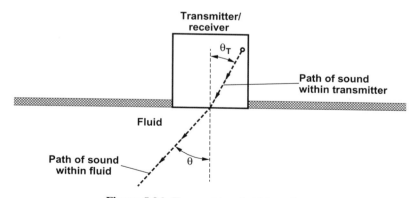

Figure 5.36. Transmitter–fluid interface.

The angle of incidence θ_T within the transmitter is known. The sonic velocity V_T within the transmitter is a function of transmitter temperature, which can be sensed to compensate for its effect on the sonic velocity within the transmitter.

Solving the above expression for V gives a meter equation in which the mean fluid velocity V is linearly related to the frequency shift Δf:

$$V = \frac{\Delta f V_T}{2 f_T \sin \theta_T}$$

Boundary Layer

For Doppler flow meters, the real situation, as shown in Figure 5.37, is much more complex:

- At any stationary surface there is a boundary layer in which the fluid velocity changes from zero to the mean fluid velocity. Consequently, the frequency shift occurs gradually over this boundary layer, not instantaneously at the surface. The frequency shift in reflections from particles in the boundary layer will be less than that for particles in the main flowing stream.
- As the sonic beam travels through the fluid, it spreads and it loses intensity (it is gradually absorbed by the fluid). Both effects increase approximately with the square of the distance from the point of entry into the fluid. The reflections from particles in the boundary layer will be more intense than those from particles in the main flowing stream.

The point is that the signal from the receiver is not composed of sharp reflections from particles in the main flowing stream. Digital signal processing

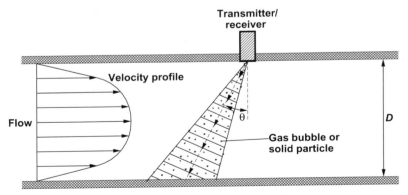

Figure 5.37. Real situation for the Doppler meter.

permits the manufacturers to incorporate routines that are better at identifying the reflections of interest. However, the situation is very complex and differs from one application to the next.

Advantages

The advantages of a Doppler ultrasonic flow meter include the following:

- Linear volumetric meter.
- Applicable to liquids, but not gases.
- No permanent pressure loss.
- No flashing or cavitation within meter.
- Low cost; easy to install.
- Clamp-on versions are noncontact and leak free.

The Doppler flow meter is a linear volumetric meter that basically senses the average velocity of the flowing stream. The capacity range of the meter is best stated in terms of mean fluid velocity, the typical limits being 0.2 to 40 ft/sec. Commercial products are designed for liquids that contain sufficient solid particles or entrained gas to provide the sonic reflection; most do not measure bidirectional flow.

Except for those with insertion probes, there is no obstruction to flow within a Doppler flow meter. There is no permanent pressure loss; there are no regions of low pressure that can lead to flashing or cavitation. Clamp-on versions do not contact the process fluid, so the effect of corrosion and other fluid characteristics on the meter components is not an issue. In toxic applications, clamp-on versions can be serviced with no risk of exposing personnel to the process fluid.

A Doppler flow meter is very cost effective in large line sizes. In clamp-on versions, the same transmitter and receiver unit is used for line sizes from 0.5 in up. Its installation involves little more than properly preparing the surface on which the transmitter/receiver is mounted.

Disadvantages

The disadvantages of a Doppler ultrasonic flow meter include the following:

- Requires particles or gas bubbles to reflect sound.
- Typical accuracy is ±2.0% to ±5.0%.
- Temperature limit is 120 °C (occasionally higher).
- Poor track record.

Doppler flow meters require that the fluid contain sufficient solid particles or entrained gas bubbles to reflect sound. Manufacturers do not generally make recommendations on straight pipe lengths upstream and downstream. However, these meters essentially assume that the reflections are from particles within the main flowing stream. In practice, how can one be sure that this is consistently the case?

Statements on the accuracy of a Doppler flow meter are generally in the ±2.0% to ±5.0% range. Some accuracy statements are based on reading; others are based on upper-range value. Turndown ratios are not generally stated.

The track record of Doppler flow meters is very poor. But on occasions, one will prove successful in a very difficult application (sometimes they become the measure of last resort). Their low cost and ease of installation make it very tempting to give them a try. However, one should resist the temptation to merely throw technology at a problem.

5.13. THERMAL FLOW METERS

Although theoretically applicable to measuring liquid flows, most industrial applications of thermal flow meters are for measuring gas flows. There are two types of thermal flow meters:

Rate of heat loss. A heated element is at a temperature higher than the temperature of the flowing stream. Let ΔT be the difference between the temperature of the element and the fluid temperature. Let q be the rate of heat loss from the element. There are two approaches:

- Constant power. At a fixed heat transfer rate q, the temperature rise ΔT is measured. The value of ΔT decreases with the fluid mass flow rate.
- Constant temperature. For a fixed temperature rise ΔT, the heat transfer rate q is measured. The value of q increases with the fluid mass flow rate.

Temperature rise. A known amount of heat is added to a flowing stream. The upstream and downstream temperatures are measured. The temperature rise decreases as the fluid mass flow rate increases.

Rate of Heat Loss

One type of thermal flow meter is based on the theoretical equation for the rate of heat loss from a small cylinder (such as a wire) in a flowing fluid:

$$q = \left[k + 2(kc_{p}\rho \pi dV)^{1/2} \right] \Delta T$$

where q = rate of heat loss per unit length of the cylinder; k = fluid thermal conductivity; c_{p} = fluid heat capacity; ρ = fluid density; d = diameter of the cylinder; V = fluid velocity; ΔT = temperature difference between cylinder and fluid.

For a given gas, the thermal conductivity k is small, so the $k+$ term is negligible. For a given fluid, k and c_p are known. Because d is also known, the equation can be approximated (for gases) as follows:

$$q \cong K(\rho V)^{1/2} \Delta T = Km^{1/2}\Delta T$$

where $K = 2(kc_p\pi d)^{1/2}$; $m = \rho V$ = the mass flux (mass flow per unit area) $[(\text{lb}_m/\text{sec})/\text{ft}^2$ or $(\text{kg/sec})/\text{m}^2]$.

The relationship between q and ΔT depends on the mass flow in a nonlinear fashion (note $m^{1/2}$ in the expression). While these theoretical equations suggest the general relationships, in practice each thermal flow meter must be individually calibrated. The signal processing also includes the linearization necessary to provide an output signal that varies linearly with the mass flow.

Insertion Probe

Although in-line assemblies are available for pipes of 6 in or less, most installations of the rate of heat loss type of thermal flow meters use an insertion probe (Fig. 5.38). There are two temperature sensors:

Fluid temperature. Either an RTD or a thermocouple is used to sense the fluid temperature. No condensation must occur on this sensor (fluid temperature must be above its dew point).

Hot element temperature. Some versions heat this element by passing current through an RTD; others apply heat by passing an alternating

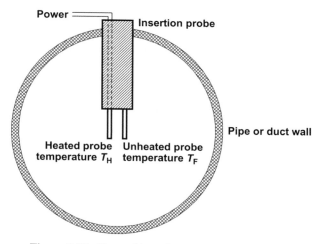

Figure 5.38. Rate of heat loss thermal flow meter.

current through a thermocouple. Both the power input and the hot element temperature must be measured. The amount of heat transferred to the flowing fluid is so small that the temperature rise of the flowing fluid is negligible and is not sensed.

In effect, the thermal flow meter senses the point mass flow at the location at which the probe is inserted into the flowing stream. Depending on the flow distribution in the pipe or duct, locating at the centerline (Fig. 5.38) might not be appropriate. Many insertion probes permit the insertion depth to be varied so that the effect on the measured mass flow can be determined.

Characteristics

The characteristics of the rate of heat loss thermal flow meter include

- Applicable to most gases.
- Particulates, condensates, and high velocities should be avoided.
- Typical accuracy is 2.0% of upper-range value.
- Turndown ratio of 100:1.
- Insertion type is easy to install.
- Temperature limit as low as 50 °C, as high as 260 °C.
- Small pressure drop (usually < 0.2 psi).
- Requires straight pipe upstream and downstream.

The rate of heat loss thermal flow meter is applicable to air, oxygen, nitrogen, chlorine, ammonia, CO, CO_2, light hydrocarbons (including natural gas), etc. However, low-density gases (helium and hydrogen) lead to low heat losses and potentially poor performance.

The issues between constant power and constant temperature include the following:

Zero flow. When maintaining a constant power input, the temperature of the heated element increases rapidly as the flow approaches zero. To prevent overheating at zero flow, the power input must be reduced. But the lower the power input, the smaller the temperature rise under flowing conditions. This degrades the performance of the meter.

Response time. When a constant power input is maintained, a change in the mass flow leads to a change in the heated element's temperature. But the thermal mass of this element degrades the response time. Maintaining a constant temperature results in a faster response to a change in mass flow.

Hot Wire Anemometer

The hot wire anemometer is a low-cost sensor commonly used to measure air velocity in large ducts, such as in heating, ventilation, and air-conditioning

(HVAC) applications. These units can sense air velocities up to about 30 ft/sec to an accuracy of ±3% of reading. Their low cost permits several sensors to be installed at various positions within a duct to obtain the flow distribution. Handheld models can be used to sense the airflow from a vent.

In its original configuration, a hot wire anemometer consisted of an exposed hot wire (tungsten or platinum) supported by two posts. The exposed wire is thin and easily damaged. Rugged designs use platinum films deposited on glass tubes or flat surfaces. Some models now use heated thermistors instead of the exposed wire. In either case, two configurations are possible:

Constant current. A constant current is applied to the wire, and its temperature is sensed. The wire temperature is a function of air velocity.

Constant temperature. The current applied to the wire is adjusted to maintain a constant wire temperature. The current flow is a function of the air velocity. This is the preferred approach.

Heated Tube

In both the insertion and the in-line thermal flow meters, the sensor elements are exposed to the fluid and consequently to corrosion, buildups, etc. The heated-tube designs address these issues by locating both the temperature sensors and the heater element external to the pipe, as shown in Figure 5.39. There are two temperature sensors:

Fluid temperature T_F. This sensor is located sufficiently far upstream that it is not affected by the heater element. Therefore, this sensor measures the fluid temperature.

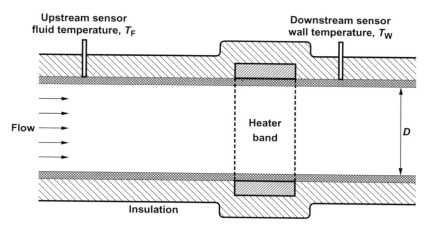

Figure 5.39. Heated tube thermal flow meter.

Wall temperature T_W. This sensor is located slightly downstream of the heater element. This sensor measures the wall temperature, which is the driving force for heat transfer to the fluid.

The heater element is insulated so that a negligible amount of heat is lost to the surroundings. At equilibrium, the power supplied to the heater element must be the same as the rate of heat transfer q to the fluid. The heat transfer equation is

$$q = hA(T_W - T_F)$$

where h = heat transfer coefficient; A = heat transfer area.

For turbulent flow in circular pipes, the Dittus-Boelter equation relates the heat transfer coefficient h to the pipe geometry, the fluid properties, and the mass flow:

$$N_{Nu} = 0.0243 N_{Re}^{0.8} N_{Pr}^{0.4} (\mu/\mu_W)^{0.14}$$

where N_{Nu} = nusselt number = hD/k; N_{Re} = Reynolds number = $4W/(\pi D\mu)$; N_{Pr} = Prandtl number = $c_p\mu/k$; h = heat transfer coefficient; D = pipe diameter; k = fluid thermal conductivity; W = mass flow; c_p = fluid heat capacity; μ = fluid viscosity at T_F; μ_W = fluid viscosity at T_W.

The pipe geometry and the fluid properties are known. Lumping these into a coefficient K permits the Dittus-Boelter equation to be written as follows:

$$h = KW^{0.8}$$

Combining with the heat transfer equation gives the meter equation for a heated tube thermal flow meter:

$$W^{0.8} = \frac{A(T_W - T_F)}{Kq} = \frac{K'(T_W - T_F)}{q}$$

This expression gives the general form of the relationship, but one must proceed with caution. The expression is for turbulent flow; a similar expression can be derived for laminar flow, except that the exponent on W is 0.33. In practice, a heated tube thermal flow meter must be calibrated for the process conditions at which the measurement is to be made.

Temperature Rise

As illustrated in Figure 5.40, the temperature rise thermal flow meter consists of three elements:

- Heater or cooler that adds or removes a known or measurable quantity of heat to or from a flowing stream.

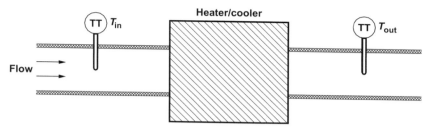

Figure 5.40. Temperature rise thermal flow meter.

- Temperature sensor to measure the fluid temperature T_{in} upstream of the heater or cooler.
- Temperature sensor to measure the fluid temperature T_{out} downstream of the heater or cooler.

Commercial versions of these flow meters are available only for measuring small gas flows. However, they can potentially be applied anywhere heat is being added or removed from a flowing stream. If the amount of heat being added or removed can be measured, the mass flow can be calculated from the heat flow and the temperatures in and out.

Although Figure 5.40 suggests the individual temperatures are measured, only the temperature difference will be required in the meter equation. Sensing the temperature difference directly generally provides a more accurate value than sensing the individual temperatures and computing the difference.

Temperature Rise Meter Equation

To obtain the meter equation for a temperature rise thermal flow meter, the sensible heat equation is solved for the mass flow W:

$$q = Wc_p(T_{out} - T_{in}) = Wc_p\Delta T$$

$$W = \frac{q}{c_p\Delta T}$$

Note that q and ΔT are sensed, and W is computed. The only remaining parameter is the fluid heat capacity c_p. This parameter is a function of

Fluid temperature. Relationships giving the effect of temperature on the heat capacity are available for most components. If the fluid composition is known, the measurement can be compensated for variations in fluid temperature.

Fluid composition. Compensating for variations in fluid composition requires the ability to measure the fluid composition. Unfortunately, analyzers tend to be expensive, in terms of both initial cost and maintenance.

When a fixed quantity of heat is added to the flowing stream, a temperature rise thermal flow meter will have a turndown ratio of perhaps 10:1. As the flow increases, the temperature change across the exchanger decreases. The difficulty in accurately sensing small temperature differences generally imposes limits on the measurement range that a given flow meter can handle. But in many applications, the heat flow increases as the mass flow increases (the temperature rise is constant or nearly constant). In this case, the turndown ratio is limited only by the measurement range for the heat flow.

5.14. TURBINE METER

Figure 5.41 provides a simplified representation of an axial turbine meter. The major components are

Rotor. This consists of a hub and blades. In small turbine meters, the blades usually have a fixed pitch. In larger meters designed for gas service, the blades are usually slightly curved, having a shape somewhat like an airplane wing.

Bearing assembly. This consists of a shaft (extending through the rotor hub) and two bearings, one fore and one aft (more on bearings later).

Diffusers. The diffusers, one upstream and one downstream, provide mounts for the rotor bearings. The diffusers cause the fluid velocity to increase (and the pressure to decrease). The diffusers are aerodynamically shaped so as to minimize turbulence and enhance pressure recovery.

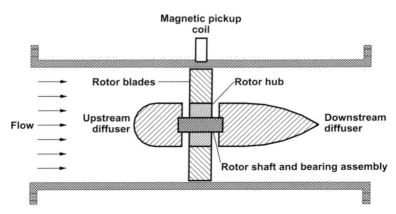

Figure 5.41. Turbine meter.

Diffuser supports. Although not shown in the figure, supports are required for both the upstream and the downstream diffusers. The supports must not introduce turbulence, so they usually resemble straightening vanes. However, separate straightening vanes are usually installed at an appropriate location upstream of the turbine meter.

Meter Factor

Rotation is usually detected by a magnetic pickup coil, although mechanical detectors and RF detectors are sometimes used. The disadvantages of mechanical detectors should be obvious. Magnetic detectors provide some drag on rotation, which increases in significance as the flow decreases. The RF frequency detectors contain a radio frequency oscillator. When an electrically conducting rotor blade passes in proximity to the detector, the impedance changes and is sensed by the detector. This provides no drag on the rotor, but RF detectors are more expensive than the magnetic types.

Basically, the detector is digital. Most designs detect each rotor blade as it passes under the detector. But in designs with nonmagnetic blades, either a permanent magnet or some type of magnetic material must be embedded in the rotor. These designs detect each rotation of the rotor. In either form, the output is a frequency signal, which is easily shaped to generate a sequence of pulses. A frequency-to-current converter is required to obtain a 4- to 20-ma signal.

To determine the flow through the meter, one needs to know the volume of fluid that corresponds to each pulse from the detector. Most manufacturers state the reciprocal. The meter factor (usually designated by K) is usually stated as the number of pulses per unit volume of fluid—that is, pulses per gallon, pulses per liter, pulses per cubic foot, and pulses per cubic meter. Linearity expresses the flow range over which K is constant. At low flows, K initially increases slightly, but then decreases very rapidly.

Bearings

A turbine meter has only one moving part: the rotor. This requires bearings. There are two types of bearings commonly used in turbine meters:

Sleeve bearings. These are generally recommended for liquid service, provided the liquid provides some degree of lubrication. Tungsten carbide is most commonly used, but there are other possibilities.

Ball bearings. These are generally recommended for gas service and for nonlubricating liquids.

If one is experiencing problems with a turbine meter, the bearings are almost always the culprit. Regardless of the type of bearing, they are immersed in the fluid and thus are exposed to corrosion, particulate matter, etc. Turbine meters are recommended only for clean fluids but are occasionally successful in other applications. Involvement from the turbine meter manufacturer before purchase is crucial. Try to find one with experience in your specific fluid. If others have not been successful, your prospects are not good. But even when others have been successful, your success is not ensured. Even trace amounts of a problem impurity can have an effect on the bearings.

Excess Velocities

Turbine meter bearings are damaged by excess rotational speeds. One definite no-no: never blow air (or other gas) through a turbine meter designed for liquid service. Unfortunately, there are process situations in which exactly this can happen.

The process in Figure 5.42 is a typical feed system for a batch reactor. There are four feed materials. Depending on the product, they are charged in differ- ent amounts and in a different order. However, they are never charged simul- taneously. This permits a single flow meter to be shared between these four feeds.

In such applications, issues often arise regarding mixing feed materials before they enter the reactor. Reactions must never occur in the pipes that deliver materials to vessels—the consequences are invariably bad. The stan- dard operating procedure is to transfer one material, then blow the line dry with an appropriate gas, and then transfer another material. This procedure will damage a turbine meter. While the line is being blown dry, the rotor spins too fast, and without any liquid to lubricate the bearings. Some other alterna- tive, such as purging the feed lines with a nonreactive liquid (such as a solvent), must be used to avoid mixing inappropriate materials.

Pressure Issues

Most turbine meter designs are proprietary, so here we can provide only general statements. However, turbine meter manufacturers are very good at

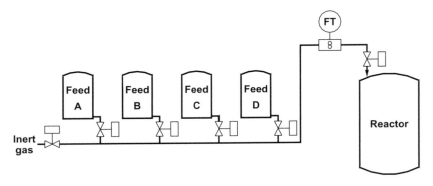

Figure 5.42. Batch reactor feed system.

providing data specific to their products. There are two issues pertaining to pressure:

Permanent pressure loss. The permanent pressure loss must be determined using the graphs or tables provided by the turbine meter manufacturers for their products. For liquids, the permanent pressure loss at maximum rated capacity is usually less than 10 psi. For gases, the permanent pressure loss is usually less than 0.1 psi.

Point of lowest pressure. The diffusers increase the velocity and reduce the pressure. The point of lowest pressure is usually just upstream of the rotor. In liquid service, the potential for flashing and cavitation must be considered. Minor flashing merely causes the meter to read high (gas is much less dense than liquid). In extreme cases, flashing leads to excessive rotor speeds and mechanical damage. To avoid flashing, the minimum process pressure at turbine meter inlet is usually computed by the following equation:

$$P > k_1 P^\circ + k_2 \Delta P$$

where P = process pressure at meter inlet; P° = fluid vapor pressure; ΔP = permanent pressure loss; k_1, k_2 = coefficients provided by meter manufacturer.

Typical values are $k_1 = 1.5$ and $k_2 = 2.0$, but these depend on the design of the turbine meter. The recommendations from the meter manufacturer must be followed. In marginal cases, a different meter size can be used to reduce the potential for flashing, but with some reduction in measurement range.

Advantages

The advantages of a turbine meter include the following:

- Applicable to gases and liquids.
- Very accurate (±0.1%) linear volumetric meter.
- Turndown ratio up to 100:1.
- Small to modest permanent pressure loss.
- Temperatures up to 450°C.
- Pressures up to 5000 psig.

Before the introduction of coriolis mass meters, turbine meters were the most accurate flow meters available. They were commonly installed in batch plants to meter chemicals into reactors. But in this application, they have been largely displaced by coriolis meters, one reason being that the coriolis meter is a mass meter, whereas the turbine meter is a volumetric meter. But in many cases, turbine meters worked quite well. You will encounter some older installations in which the turbine meters have not been replaced.

Today, the most common application for turbine meters is measuring gas flows. The line sizes are often large, which makes coriolis meters prohibitively expensive. The pressure drop from a turbine meter will be much less than from a coriolis meter. By incorporating temperature and pressure compensation, the volumetric flow can be accurately converted to a mass flow.

Pressure ratings are normally determined by marketing considerations; if required, a turbine meter could be designed for any pressure rating. Temperature limitations are imposed by the electrical components that detect rotor rotation and by issues pertaining to the bearings.

Disadvantages

The disadvantages of a turbine meter include the following:

- Recommended only for clean, low-viscosity fluids.
- Requires straight pipe upstream and downstream.
- Bearing selection is crucial.
- Potential for flashing and cavitation.

The main concern with every turbine meter installation pertains to the bearings. If problems are experienced, the bearings are invariably involved. The manufacturers have introduced various features into their designs to address the problems with the bearings, but nothing seems to be infallible. Manufacturers often specify a maximum size for the particles permitted within the fluid. But bearings are subject to chemical attack as well as deposits from the fluid.

Turbine meters are sensitive to the velocity profile of the fluid as it enters the rotor. The recommendations include appropriate lengths of straight pipe upstream plus vanes at the proper location. Because accuracy is one of the appealing attributes of a turbine meter, the required straight lengths of pipe and vanes must be provided in most applications.

In liquid applications, high viscosities degrade the performance of turbine meters. If the viscosity of the fluid is greater than 5 cSt, one should proceed with caution. In gas applications, low densities (from gases such as hydrogen and helium, or from low pressures) also degrade the performance of turbine meters.

5.15. OTHER FLOW METERS

The use of positive displacement meters, rotameters, and target meters in industrial applications will be discussed briefly.

Positive Displacement Meters

Probably the most familiar application of such meters is residential water and gas, both of which are very clean fluids. The key requirements of this application are

- No external source of power is required.
- Only the flow total is required; the instantaneous flow rate is of no interest.

Industrial applications with such requirements are good candidates for positive displacement meters, but there are limitations:

1. Adding a remote readout is certainly possible. Reading the flow total is complicated by the mechanical nature of the positive displacement meter. The simplest interface is to incorporate a contact that closes (or opens) momentarily when the least significant digit on the totalizer changes by one unit, thus providing a signal consisting of a sequence of pulses.
2. Determining the fluid flow rate is a challenge. The time interval between the pulses varies from a fraction of a second to several seconds. How do you convert such a slow sequence of pulses into a flow rate? Counting pulses takes too long. The time interval between pulses is a possibility, but there is usually some variability in this time interval.

The following statements summarize the issues pertaining to the use of positive displacement meters in industrial applications:

- Applicable for gases and clean liquids.
- Accuracy is ±0.25%.
- Generally accepted for custody transfer.
- No external source of power.
- Unaffected by upstream piping.
- Mechanical wear degrades accuracy.
- Expensive in large sizes or special materials.

Rotameters

The generic term for rotameters is *variable area flow meter*. As shown in Figure 5.43, fluid (liquid or gas) flows upward through a tapered tube containing a float. The clearance between the float and the tapered tube forms an orifice through which the fluid must flow. This results in a pressure drop across the float, which is one of three forces acting on the float:

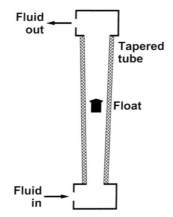

Figure 5.43. Variable area flow meter.

- Force due to gravity (weight of the float).
- Buoyant force (significant only in liquid applications).
- Pressure drop across the float (produces an upward force on the float).

The upward force resulting from the pressure drop must equal the weight of the float less any buoyant force. If the liquid specific gravity is constant, the weight of the float less the buoyant effect is also constant. Consequently, the pressure drop across the float must be constant. For a given fluid flow, a certain orifice size is required to give this pressure drop. The float position at equilibrium will be where the clearances between the float and the tapered tube provide an orifice of the required size.

In past years, rotameters were commonly used to measure process flows. But this is becoming less frequent. The most common use of rotameters is as flow indicators in equipment such as analyzer sample systems. The designers of analyzer sample systems stress the importance of knowing the sample flows. Rotameters are good inexpensive indicators for such small flows. For this application, a needle valve is often incorporated into the rotameter assembly so that the flow can be adjusted when necessary.

The glass tube in older rotameters has been replaced by a transparent plastic tube (options include PVC, polyamide, polysulfone, PVDF). With magnetic detectors, a remote readout and/or low–high flow limit switches are possible. Also, a metal tube can be used instead of the plastic tube. However, neither is common industrial practice. Rotameters are basically local flow indicators.

Rotameters can be used only for clean fluids. Deposits on the float affect accuracy (typically ±1 to ±4% of full scale); entrained gas bubbles can also adhere to the float and affect accuracy. Indicator scales usually apply to either water or air at standard conditions. For other fluids, scale correction factors

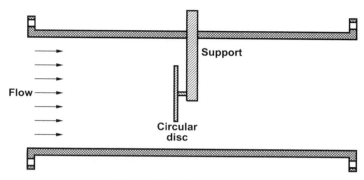

Figure 5.44. Target meter.

based on specific gravity must be applied. For gases, the effect of pressure and temperature is also incorporated via scale correction factors.

Target Meters

A target meter inserts an obstruction to flow, such as the circular disk shown in Figure 5.44, into the flowing stream. A mechanism is required to sense the force exerted on the obstruction. This force is due to two factors:

- The impingement of the flowing fluid onto the obstruction.
- The viscous drag of the fluid flowing around the obstruction.

The target meter is nonlinear (the force increases approximately with the square of the fluid flow), and its accuracy is difficult to establish. Especially for liquid applications, changes in fluid properties affect the meter. Abrupt changes can occur, but in most cases the changes occur rather slowly. Consequently, the repeatability, at least in the short term, is reasonably good. In short, target meters should be considered for applications requiring good repeatability but not necessarily good accuracy.

Target meters are normally encountered in very difficult flow measurement applications in which nothing else would work. An example is measuring the flow of a viscous material at a high temperature. These are routinely encountered in polymer processing and occasionally in refinery processes involving the bottom of the barrel.

5.16. FLOW SWITCHES

A flow switch provides a discrete input to the logic for equipment protection, detection of abnormal process events, etc. The following list provides only a few examples:

- Low flow protection for pumps or blowers.
- Loss of lubrication protection.
- Detection of loss of a chemical feed.
- Detection of flow from a relief device.

Some flow switches provide an adjustable set point; in others, the switch point is determined by modifying a mechanical part. But one characteristic remains the same: Flow switches cannot be set to change states at a precise flow. Flow switches are reasonably good at detection of low flows, which is indeed the requirement in many applications. However, if the switch is to occur at a precise flow rate, consider installing a continuous flow measurement that can be calibrated, and then compare the measured value against the limits to detect either low flow or high flow.

Paddle

The paddle flow switch illustrated in Figure 5.45 is a one version of a mechanical flow switch. The flowing fluid impinges onto the paddle, causing it to deflect in the direction of flow. To eliminate a possible source of leaks, the paddle is usually magnetically coupled to a switch that is actuated at a certain deflection of the paddle.

These are inexpensive but relatively crude devices. The length of the paddle is modified depending on the line size (1 in or larger) and the flow required to

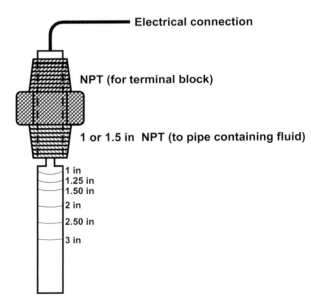

Figure 5.45. Paddle flow switch.

deflect the switch (the longer the paddle, the lower the flow required to deflect the switch). These devices are obviously affected by deposits and buildups, which can cause the paddle to "hang" in the deflected position.

These are applicable to both liquids and gases. Paddle flow switches do not require an external source of power.

Thermal Dispersion

The thermal dispersion flow switch shown in Figure 5.46 is based on the same principles as the thermal mass flow meters previously described. The sensor consists of two temperature probes (RTDs), one heated and one not heated. As the flow decreases, the temperature of the heated sensor increases, and quite rapidly at low flows.

The set point to the thermal flow switch is the temperature differential at which the switch is to actuate. The thermal flow switch has an adjustable set point; the problem is determining the temperature difference that corresponds to the flow at which actuation is to occur.

These are applicable to both liquids and gases. Probes are available with immersion depths of less than 1 in (for pipe sizes from 0.75 to 1.50 in) and 3 in (for larger pipes). Plastic construction of the probe imposes temperature limits of 32° to 140 °F and a pressure limit of 150 psi or less, depending on temperature. Metal construction of the probe provides temperature limits of −100° to 850 °F and a pressure limit of 1500 psig (or higher for some constructions).

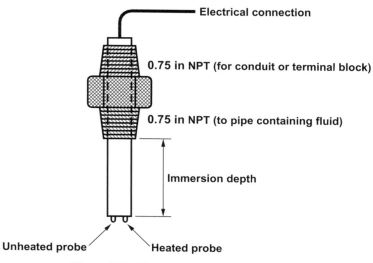

Figure 5.46. Thermal dispersion flow switch.

Doppler

The Doppler flow switch is based on the same principles as the Doppler flow meters described previously. The ultrasonic transmitter directs sound into the flowing stream at an angle, and the receiver listens for the reflection from either solid particles or gas bubbles within the flowing stream. The Doppler shift decreases as the flow decreases. The transmitter and receiver unit and signal processing are the basically the same as in the Doppler flow meters, making the Doppler flow switches expensive compared to other flow switches. The set point for Doppler flow switches is adjustable.

The installation of the transmitter and receiver unit is external to the pipe wall, so a Doppler flow switch is noncontact. The transmitter and receiver are within a single element that is attached to the outer surface of the pipe with a bonding agent, band clamps, etc.

Doppler flow switches are applicable only to liquids. Furthermore, they depend on the presence of either solid particles or gas bubbles in the flowing stream to serve as reflectors of sound.

LITERATURE CITED

1. Yaws, Carl L., *Chemical Properties Handbook*, McGraw-Hill, New York, 1999, table 7.
2. International Organization for Standardization, *Measurement of fluid flow by means of pressure differential devices inserted in circular cross-section conduits running full—Part 1: General principles and requirements*, ISO 5167-1, ISO, Geneva, 2003.
3. International Organization for Standardization, *Measurement of fluid flow by means of pressure differential devices inserted in circular cross-section conduits running full—Part 2: Orifice plates*, ISO 5167-2, ISO, Geneva, 2003.
4. International Organization for Standardization, *Measurement of fluid flow by means of pressure differential devices inserted in circular cross-section conduits running full—Part 4: Venturi tubes*, ISO 5167-4, ISO, Geneva, 2003.

Basic Process Measurements, by Cecil L. Smith
Copyright © 2009 by John Wiley & Sons, Inc.